The Cambridge Manuals of Science and
Literature

PEARLS

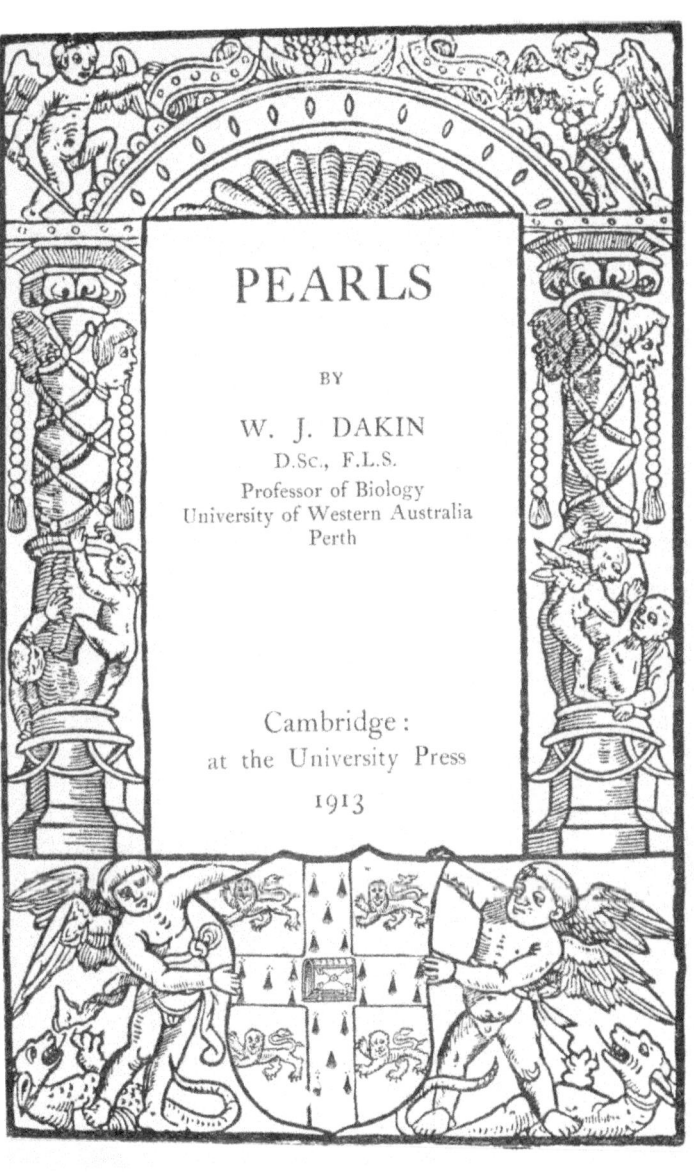

PEARLS

BY

W. J. DAKIN
D.Sc., F.L.S.
Professor of Biology
University of Western Australia
Perth

Cambridge:
at the University Press
1913

CAMBRIDGE UNIVERSITY PRESS
Cambridge, New York, Melbourne, Madrid, Cape Town,
Singapore, São Paulo, Delhi, Tokyo, Mexico City

Cambridge University Press
The Edinburgh Building, Cambridge CB2 8RU, UK

Published in the United States of America by Cambridge University Press, New York

www.cambridge.org
Information on this title: www.cambridge.org/9781107616622

© Cambridge University Press 1913

First published 1913
First paperback edition 2011

A catalogue record for this publication is available from the British Library

ISBN 978-1-107-61662-2 Paperback

Cambridge University Press has no responsibility for the persistence or
accuracy of URLs for external or third-party internet websites referred to in
this publication, and does not guarantee that any content on such websites is,
or will remain, accurate or appropriate.

*With the exception of the coat of arms at
the foot, the design on the title page is a
reproduction of one used by the earliest known
Cambridge printer, John Siberch,* 1521

PREFACE

NO one can be better aware of the defects in this little book than the author. It has been written with a view to giving a summary of the most important facts about pearls, pearl fishing and pearl formation, and it is hoped by the writer that any sins of omission will be pardoned. There is no lack of literature on the subject of pearls, and some of the volumes which have been published on this subject are themselves of great value. They do not seem, however, to be easily within reach of the average reader, and they are nearly always written from non-biological points of view. The works on pearl formation are, on the other hand, usually very technical. The structure and natural history of the shell-fish in which pearls are produced are hardly ever described outside scientific treatises; hence the chapter on this subject. It seemed impossible, to the author, to describe the processes of pearl formation without reference to the structure of the shell-fish. If it were only for common knowledge, a description of the anatomy of a mollusc should be interesting to the general reader, for very little idea of the structure of even an edible oyster or mussel seems to prevail. The book has been written by a biologist, from the point of view of a biologist, and owes much to previous writers who have treated the subject from other sides. To Professor Herdman, my former teacher, I owe very much, and his descriptions of the Ceylon banks have been very largely utilised. The writing up has only been accomplished through the assistance rendered by my wife, for by a curious coincidence it had to be written under the stress and strain of preparations for a removal to one of the countries referred to on many occasions in the text, Western Australia. Mr H. Jackson, M.Sc., has very kindly undertaken the labour of proof reading; without this help considerable delay would have occurred in the publication. To Mr Chambers, of University College, London, I owe several interesting references to pearls in the classics; and Dr Johnstone, of Liverpool, has read certain parts of the manuscript.

If I have occasionally fallen into the pitfall of becoming too technical, I trust that my readers will turn to the glossary, and I earnestly hope that it will help to clear up the difficulty.

W. J. D.

R.M.S. *Mongolia*,
    Mediterranean.
        February 1913.

CONTENTS

CHAP.		PAGE
I.	THE HISTORY OF PEARLS.	1
II.	WHERE PEARLS ARE FOUND	11
III.	THE ANATOMY OF THE PEARL OYSTER AND THE STRUCTURE OF ITS SHELL	17
IV.	THE LIFE HISTORY OF THE PEARL OYSTER, AND THE CONDITIONS UNDER WHICH IT LIVES	42
V.	PEARLS	52
VI.	A CEYLON PEARL FISHERY	68
VII.	PEARL FISHERIES OF OTHER LANDS	83
VIII.	THE ORIGIN OF PEARLS	91
IX.	PEARLS IN JEWELLERY AND TRADE, ODD FACTS	116
X.	THE PEARL OYSTER, THE TRADER, AND THE SCIENTIST	124
	BIBLIOGRAPHY	138
	GLOSSARY	140
	INDEX	144

LIST OF ILLUSTRATIONS

FIG.		PAGE
1.	Valve of Ceylon pearl oyster	19
2.	Section (diagrammatic) of molluscan shell	21
3.	Nacre laminae in surface view	23
4.	Anatomy of Ceylon pearl oyster	28
5.	Positions of muscle pearls in Ceylon pearl oyster	55
6.	Section of pearl (prismatic)	57
7.	Section of pearl (nacreous)	59
8.	Nemertine worm imbedded in nacre	86
9.	Diagrams showing formation of pearl in Mytilus	103
10.	Section of larval worm in Ceylon pearl oyster	105
11.	*Tetrarhynchus unionifactor*	105
12.	Diagrams of development of pearl sac	113

CHAPTER I

THE HISTORY OF PEARLS

THE use of pearls as jewels and their recognition as objects of value date back into the far beyond when the histories of ancient peoples were transcribed upon papyrus. It is very likely that pearls were amongst the earliest gems known to man, and this is not surprising when one considers that the earliest dwellers by the sea probably fed upon the shellfish which produce such objects.

The modern recognition of the pearl dates back to about 300 B.C. and was due, no doubt, to the mad desire of the Romans for luxury and treasure; but although pearls seem to have played a less conspicuous part in Ancient Greece and Egypt, they were known to yet older peoples, and especially to those of the East. The Old Testament only refers to pearls once, in the book of Job (chap. xxviii), where it is written, "No mention shall be made of coral, or of pearls; for the price of wisdom is above rubies." None of the terms, however, refer with any certainty to the objects now known by such names. The names of gems in the scriptures have probably arisen during the many translations and rewritings of the books.

In the New Testament, on the other hand, constant reference is made to pearls. There are, however, traditions which refer to the ancient Hebrews having knowledge of the true value of the gem.

In the Indian Seas pearls were known many centuries before Christ. Perhaps first collected as mere curiosities or as objects of some superstitious regard, they gradually became articles of greater and greater value, and the proud possessions of the Princes of the East. References occur in the literature of India which are of considerable antiquity. Pearls are mentioned in the books of the Brahmans (about 500 B.C.) and are associated in Hindu literature with Krishna.

The Chinese records go still further back and Kunz and Stevenson relate that a book, the Shu King (2350—625 B.C.), states that in the 23rd century B.C. Yü received as tribute, oyster pearls from the river Hwai. The early Chinese pearls were evidently taken from fresh-water molluscs.

The pearl fisheries of Ceylon and also of India and the Persian Gulf, must be of very great antiquity, but when and by whom discovered, are questions the answers to which are buried in the annals of the past. It is highly probable that the fishing was carried on 2000 years ago, in much the same simple way as at present. The fishing of Ceylon was referred to by Pliny, who stated that the island of Ceylon was the most productive of pearls of all parts of the world. According to Herdman the Singhalese records go back still further, and quoting from his valuable report—a mine of information about the fishing of Ceylon—we read that "According to the 'Mahawanso,' pearls figure in the list of native products sent as a present from King Vigáya of Ceylon to his Indian father-in-law, in about 540—550 B.C."; and again when in B.C. 306 King Devanampiyabissa sent an embassy to India the presents are said to have included eight kinds of Ceylon pearls.

Pearls were almost certainly known to the Persians seven centuries before Christ; they are not mentioned, but pearl ornaments of very great antiquity have been found in Persian remains.

The ancient Egyptians used mother of pearl, according to Kunz and Stevenson, as early as the 6th dynasty—3200 B.C., but they do not seem to have regarded the pearl as being valuable until very much later times.

As for the Greeks, they too knew the pearl and recognised its value. The gems are mentioned in the writings of Theophrastus and they are described as being the products of shellfish. Pliny also refers to the writings of the Greeks. Three centuries before Christ a great change in the story seems to have taken place. The Roman empire was beginning to rise and very soon this mighty power became involved in wars with those countries where the knowledge of the pearl was general and its value as treasure was recognised. By 50 B.C. the pearl had become very popular in Rome. The Romans were indeed strangely affected by pearls, and the gem was adopted as a kind of fetish—a sign of pomp and luxury. The value of the pearl became extraordinarily great and laws were made forbidding the wearing of pearls by individuals who had not attained to a certain rank. Dresses were simply covered with the gem and even animals wore necklaces. It is often said that the presence of pearls in the molluscs of the rivers of the British Islands, played some part in bringing Julius Caesar to our shores. It is certain that in those days the pearl was fished in England and in this respect the following quotations are interesting :

From Caius Plinius Secundus.

In Britain, it is certain that small and badly coloured pearls are found, since the Emperor Julius wished it to be understood that the breastplate which he dedicated to Venus Genetrix in her temple was made of British pearls. Lib. IX. 557.

From Caius Julius Solinus.

Dat et India margaritas, dat et litus Britannicum; sicut divus Julius thoracem, quem Veneri Genetrici in templum ejus dicavit, ex Britannicis margaritis factum, subjecta inscriptione testatus est. *C. L.* iii.

From Tacitus.

Britain bears gold, silver and other metals as the reward of victory, the ocean also produces pearls, but dull coloured and dirty brown. C. C. Taciti *de vita Agricolae* c. 5.

From Aelian.

The best [pearl] is the Indian, and that of the Red Sea. It is produced also in the western ocean where lies the Brettanic island; but appears in a measure rather golden coloured having rays somewhat dull and dusky. Juba says it is found also in the Gulf near the Bosphorus, but that it is inferior to the Brettanic.

Aeliani *de Natura Animalium* lib. xv. c. 8.

From Origen.

But they hold the second rank, as among pearls, do those which are taken from the ocean near Britain. They say that the pearl obtained near Britain is golden coloured on its surface but cloudy and rather dull in its rays; but that which is gotten from the Bay near the Bosphorus is more cloudy than the Brettanic.

Orig. *Comment. in Mattheum*, Delarue t. II. 448—50.

Most of the above writers seem to have regarded the British pearl as inferior to the pearls of the Orient. The Venerable Bede speaks more highly of the English pearls, and writes:

"—among which are mussels in which they often find enclosed pearls of all the best colours—that is, both red and purple, jacynth and green, but principally white."

So did Britain in these times play its part in the production of gems, which the makers of history regarded as of first importance.

The Romans used the name *margaritae* (of the Greeks), for pearls, but the term *Unio* was also common in Rome. Pliny explains the term Unio as meaning that each pearl was unique, but other theories have been put forward to explain this name.

After the fall of Rome and the scattering of its treasures, we find that pearls once more become objects of great value with the rise of another conquering race. Byzantium or Constantinople became the centre and capital of this new "Empire of the East," and with the development of life and luxury arose the desire for adornment which was even more gorgeous than that of the Romans. The treasures of Rome however had been scattered far and wide and the pearl travelled far over Europe even to the ancient cities of Gaul. Then we find that as the Franks, too, became a prosperous and conquering race under Charlemagne, the pearl again came into great favour with the rich and powerful. Later, when learning became the chief object in life and books the greatest treasure, the pearl was chosen to make beautiful the bindings of these books. Many of them were most splendid and costly. One, the Ashburnham manuscript of the Four Gospels, which long ago belonged to the Abbey of Canonesses on Lake Constance, is now in the possession of Mr Pierpoint Morgan. This MS. was bound about 896—899 A.D. by order of the Emperor Arnulf of the Carolingian dynasty, and according to Kunz and Stevenson has 98 pearls on it, all of which came probably from rivers in Europe.

THE HISTORY OF PEARLS

After the 8th century, as the idea began to grow that man was the centre of the universe,—that the world was made for man,—a new use was found for pearls. It was thought that all things which grew naturally, were of direct use in helping to keep the body sound and healthy. Hence we find that many of our common herbs and lowly plants came to be used as medicine, and they were used for whatever their shape or colour suggested. So we get the origin of the "kidney" bean, and the "liver"-wort. The pearl seemed so beautiful and pure that the idea arose that it too must be of some value in this respect. So we find that the small seed pearls were used as medicine, sometimes ground into powder, sometimes swallowed whole.

It was not until about the 12th century that pearls were used in England, for the Anglo-Saxons were not artistic, in the lavish manner of more southern races. All through the 13th and 14th centuries pearls were extremely fashionable over the whole of Europe as ornaments for both men and women. They were conspicuous too in church decoration, so that from the spoil which Henry VIII obtained when many of the churches were plundered, he came into possession of numerous and costly pearls.

During the 15th and 16th centuries, pearls came into still greater favour than before. Many of the German cities revived the old restrictions of the

Romans and no maiden or married woman was allowed to wear the gem; later they were restricted to one pearl chaplet. There were many laws and regulations as to their use by knights, the most stringent being in Venice.

It was not only by the inhabitants of the Old World that pearls were discovered, for Columbus found that pearl fishing was carried on in the Gulf of Mexico, and quantities of pearls have been found in the Indian mounds, either loose or strung for necklaces and wristlets. Some were mounted in quaint and primitive fashion, all showing that in the days of swarming game and roving tribes of untrammelled savages, their queens wore pearls even as they are now worn by their fair successors. The old Spanish traders obtained pearls, by fair means and foul, from the ancient treasures of the Aztec kings. America was even known in Cadiz as the "Land of Pearls," and the gem is still fished in the Caribbean Sea, along the West Coast of Central America, and on the Pearl Islands in the Bay of Panama.

It is surprising how little was known in the very ancient days about the pearl-producing shellfish, when the pearl itself was known so well. The gems were thought by the Incas to be the "eggs" of the animal. In the time of Alexander, a writer of Mytilene in the island of Lesbos, says "In the Indian sea, off the coasts of Armenia, Persia, Susiana and Babylonia, a fish like an oyster is caught, from the flesh of which men pick out white *bones*, called by them 'pearls.'"

In Britain the most important pearl fishing was carried on in Scotland. As early as 1355 A.D. Scotch pearls are referred to in a Statute of the goldsmiths of Paris. In the reign of Charles II the pearl trade was sufficiently important to attract the attention of Parliament, and for many years a large number of pearls were "cultivated."

In 1705, we find a pearl merchant named John Spruel saying that he had been dealing in pearls for forty years but could never sell Scotch pearls in Scotland, he could only sell oriental specimens. The latter were always preferred although he could show much harder and better Scotch specimens. (This appears in an account current betwixt Scotland and England, Edinburgh 1705.) In 1860 the Scottish pearl fishery was revived by a German named Moritz Unger. He visited Scotland and bought pearls from the peasants. This resulted in a vigorous search for the gems, and in 1865 we find that the total value of the pearls sold was about £1,200. This price, however, was not maintained, as the rivers were over-fished. Now, pearls are only found irregularly in the Spey, Tay, and South Esk, and to a less extent in the Doon, Dee, Don and Forth.

Pearls have also been found in Wales, chiefly in molluscs taken from the river Conway. One of these Conway pearls was given to the queen of Charles II by her chamberlain, Sir Richard Wynn. The specimen is now believed to occupy a place in the British regalia.

During the whole of the 18th century pearls were somewhat scarce, both the Ceylon and Red Sea fisheries being unproductive. The most plentiful supplies came from the Persian Gulf and from freshwater shellfish. At this period, however, diamonds became fashionable owing to the discovery of new methods of cutting and preparing them. In spite of this rival, the pearl still continued in favour and by the end of the 19th century it was more sought after and more valuable than ever.

During this century there were discoveries of pearl banks off newly settled countries, as for example, Australia, and in this country they became one of the sources of wealth to the colonists. The growing value of the pearl is not due therefore to a decrease in the source of supply but to an ever-increasing demand for a gem which never seems out of place for personal adornment or in any scheme of decoration.

CHAPTER II

WHERE PEARLS ARE FOUND

PEARLS are found in the soft parts of aquatic animals known scientifically as molluscs, and popularly as shellfish. Inasmuch, however, as our English Board of Agriculture and Fisheries lists together such creatures as shrimps, periwinkles, crabs, lobsters and oysters as shellfish, it will be advisable, for our purpose, to use the much more correct term, Mollusca. The molluscs are known to everyone by their shells, but to very few by the much more important part—the animal that made the shell. The shell is merely an external covering, a secretion of the outer layer of the animal's body— of the animal's "skin," to speak popularly. Now if we were to try and enumerate the molluscs in which pearls occur we should probably have to give the names of almost all the bivalves. There is no reason why the sequence of events which produce pearls in one species should not occur, if not regularly, at odd intervals in other species. Not all pearls however are valuable. Most of these structures are dull and worthless. We shall therefore turn our attention to those molluscs in which are found pearls having some resemblance to the gems of that name, or which have played some important part in the scientific investigation of pearl formation.

First and foremost mention must be made of the Ceylon Pearl Oyster. The name oyster, however, is here altogether a misnomer. The pearl oyster is more closely related to our common edible mussel than to our edible oyster. The Ceylon pearl oyster (*Margaritifera vulgaris*) is not very large, the adult of about three and a half years of age measuring, on an average, about 7 by 6·5 centimetres.

Margaritifera vulgaris occurs, in addition to the Ceylonese waters, in the Persian Gulf, the Red Sea, near the Maldive Islands (where however it does not appear to be fished for pearls), East Africa, and the Malay Peninsula. It is interesting to note that since the opening of the Suez Canal this pearl oyster has migrated into the Mediterranean from the Red Sea. The pearls found in this species vary somewhat in beauty according to the district in which the mollusc occurs.

Next to the Ceylon pearl oyster, mention must be made of other species of the same marine genus *Margaritifera* which are fished for pearls, or for their shells, if these are valuable enough for the mother-of-pearl industry.

The most valuable, and at the same time the largest, mother-of-pearl oyster, is *Margaritifera maxima*, Jameson, which occurs on the north coast of Australia. Its distribution extends also down the west and east coasts for some distance. These shells are the "Port Darwin" shells, the "Queensland" shells, and "West Australia" shells of commerce. This species is also found off the coasts of New Guinea (where it is called the "New Guinea" shell), in the Malay Archipelago, and at the Philippines; these Philippine shells being known as "Manila" shells.

Another very important species is *Margaritifera margaritifera* which occurs with several varieties in the trade. One variety is found off the north coast of Australia with extensions down the east and west coasts of that continent. This is the variety known as the Australian "Black-lip." It also occurs in the Malay Archipelago with extensions to Ceylon and the Maldives. Another variety is fished at Zanzibar and Madagascar, and is known for trade purposes as the "Zanzibar shell." The variety *persica*, fished in the Persian Gulf and shipped from Bombay, is the "Bombay shell" of commerce. The variety *erythraeensis*, Jameson, is the "Egyptian shell" of the trade, and is fished at Aden and other places. From Panama comes the "Panama" shell, which is the variety *barbata*. Other varieties include the "Polynesian shells," "Gambier shells," "Tahite Black-lip," and the "New Guinea Black-lip."

Of the many other species[1] of *Margaritifera* it will perhaps suffice to mention the "Sharks Bay shell," which is fished at Sharks Bay, Western Australia, and plays an important part in the button industry.

Fresh-water molluscs in which pearls are found include many famous species belonging to the family Unionidae in the rivers of the British Islands. *Margaritana margaritifera*, Linné, occurs in the rivers of Scotland, northern England, Wales, and the mountainous parts of Ireland.

It is from this shell that the valuable fresh-water pearls have been obtained. The shell is thick, and is oblong in shape with both ends obtusely rounded. The surface is rough and dark brown in appearance. It is between five and six inches long.

The Romans came to Britain for the *Unio* pearls and there is some evidence to show that they fished the Welsh rivers for them. Many valuable pearls have also been taken from *Unio* in Scotland, and an important Scotch pearl fishery was once in existence on the Tay in Perthshire. The Irish pearl mussels seem to have decreased in number—perhaps owing to the wanton destruction which has occurred owing to the search for pearls. Pearl mussels occur in the rivers of Central Europe, and America (both Canada and the United States).

[1] For a complete list reference should be made to Jameson, *Proc. Zool. Soc.*, 1901.

WHERE PEARLS ARE FOUND

Pearls occur in *Anodonta*, a large bivalve, common in many of our English ponds and lakes. The pearls from this mollusc provided material for the work of Home and Filippi (1826 and 1852), on the origin of pearls. In Chinese rivers and lakes an allied genus—*Dipsas*, is of great importance commercially, and has been cultivated in enclosed waters. It is in this shellfish that the Chinese insert little images of Buddha or some other deity, which are eventually coated over with mother of pearl.

A considerable fishery for fresh-water pearls is carried on in the rivers of the United States of America. In fact this fishery is much more important than that of Europe. It dates back some considerable time before the white people took possession of the countries of the West. One of the most recent and greatest fisheries was that in the State of Arkansas in 1897. The genera of shellfish (which are again members of the family Unionidae) include *Quadrula, Pleurobena, Lampsilis* and *Tritigonia*.

Pearls occur very frequently in the common edible mussel (*Mytilus edulis*) of our sea coasts, and especially in those from certain areas. Numbers up to twenty-six often occur in a single mussel, and Johnstone and Scott, on one occasion at Piel, obtained 390 pearls from sixty-one mussels examined. Lest anyone should imagine that a gold mine exists in the shape of pearls in our mussel beds, we must add that the greatest value of these pearls is as material for scientific investigation. The "gems" are usually dark and dirty in colour and possess little, if any, lustre. The common edible oyster has also been found to contain pearls, and stories are current of fortunate individuals who have found them when on the point of devouring this esteemed shellfish. It is probable that the part of the story dealing with the value of the pearls contains gross exaggerations!

Another interesting mollusc in which pearls have been found, and which was at one time much more valuable than at present, is *Haliotis*—the Ear Shell. It is probably familiar, in appearance if not by name, to all readers of this book. The shell occurs in large numbers in the English Channel on the rocky shores of the Channel Islands. The exterior is a very dull dirty brown—rough and coarse in texture. A series of holes perforate the shell near one side. The internal surface is lined with most beautifully coloured nacre. In fact the play of colours is probably equalled in no other shell.

This mollusc is called the "Ormer" in the Channel Islands, and the "Abalone" in America. It will be seen that the genus has a wide distribution and as a matter of fact it has been fished on the coasts of China, Japan, California, South Africa and New Zealand. The shellfish is not a bivalve. There is only one valve, and it covers the animal, which adheres to the rocks by means of a muscular foot just as its relation, the common limpet, does. The shell has been used largely in decorations, as electric light shades, etc. but does not seem much sought after at present. The inhabitants of the Channel Islands sometimes use it to decorate the exterior of their houses.

Finally, as a representative of quite another group of molluscs, that to which the cuttlefishes belong, we must add that pearls are found in the pearl Nautilus of the tropics.

This list is by no means an exhaustive one but it will serve to indicate some of the shellfish in which pearls occur and the countries in which they are to be found.

CHAPTER III

THE ANATOMY OF THE PEARL OYSTER MARGARI-
TIFERA VULGARIS, SCHUM., AND THE STRUCTURE
OF ITS SHELL

A SIMPLE account of the anatomy and biology of the Ceylon pearl oyster is interesting and necessary to any who would follow further the romance of pearl formation.

The oyster of Ceylon belongs to a group of molluscs known popularly as bivalves, but more scientifically as Lamellibranchiata. This name refers to the characteristic plate-like gills. The group includes the well-known edible oyster, the edible mussel, cockle, and almost all the other shellfish which produce pearls. The Ceylon pearl oyster is a near ally of the scallop, a lamellibranch caught in large quantities at many places round the British Islands and enjoyed as a luxury by the few who have recognised its delicate flavour. The degree of relationship between the scallops and the Ceylon pearl oyster is, as a matter of fact, much closer than that between the latter and our edible oyster.

The soft body of the animal is enclosed between two valves of shell which are hinged together along a line which is dorsal in relation to the body. The opposite edges of the valves, at the shell opening, are therefore ventral and each valve is situated laterally on the creature's body—that is, there is a right and a left valve.

Each valve is rounded in outline, but the dorsal edge or hinge is flattened and ends in wings (or auricles) in front and behind.

The shell of a four-year-old oyster measures about 9·0 cm. from the hinge-line to the opposite ventral edge and about 8·5 cm. across the valves. The shell is not thick like that of the edible oyster or

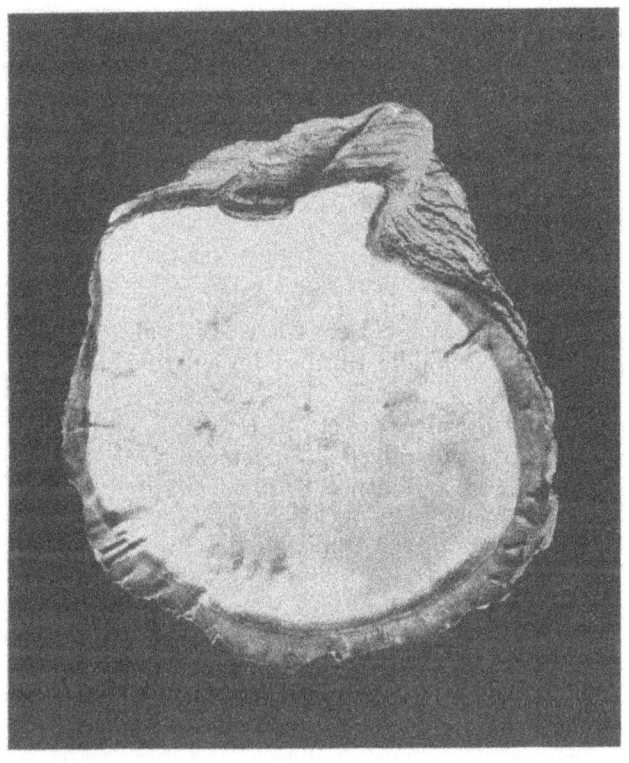

Fig. 1. Inner surface of valve of Ceylon pearl oyster (*Margaritifera maxima*).

the mother-of-pearl oysters, but very thin (only about 1·5 mm. in thickness), and it is lined by the usual nacreous layer, or mother-of-pearl, which is here exceedingly brilliant and iridescent. The mother-of-pearl layer does not extend to the edge of the valve and consequently the margin is still thinner. It is only composed of two layers, which will be referred to below. The external surface of the shell is rough and dull in colour and usually obscured by various other organisms growing upon it.

The hinge-line is a ridge which runs along the dorsal edge of each valve. Close to it and extending for about the middle third of its length is a tough black elastic ligament. This ligament, compressed when the two valves are closed, is thus always tending to divaricate the valves and open the shell. It is a dead material and consequently unaffected by the action of the animal. Inside the two valves, however, is a powerful muscle (the adductor, fig. 4) which is attached subcentrally, and this serves to close the shell. The shell valves are thus pulled together by muscular action, but they open automatically when the muscle is relaxed by virtue of the elastic ligament. For this reason the shells gape when the animal dies.

The shell of most molluscs, and practically all bivalves, like the pearl oyster and the fresh-water mussels, is composed of three layers, but a fourth occurs at places where the closing (or adductor) muscles are attached. It will be necessary to describe these layers in some detail, for the same structures are present in pearls. In fact, pearl structure is very similar to shell structure.

The outermost layer of the shell is known as the *Periostracum* (see fig. 2). It is a more or less horny layer which is normally secreted by the cells of the

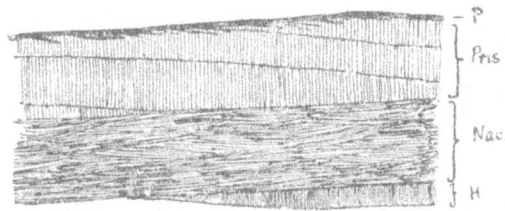

Fig. 2. Diagram of section of a typical molluscan shell. P, Periostracum. Pris., Prismatic layer. Nac., Nacreous layer. H, Hypostracum.

edge of the mantle only. Consequently it does not increase in thickness after once it has been formed. In young shellfish the periostracum is relatively much thicker than in old specimens, for the other layers increase in thickness. Very often, too, the periostracum is really thinner in the adults, because it gets rubbed off.

The periostracum is sometimes continued in the form of numerous lamellae, into the next deeper layer of shell substance. This layer is known as the *Prismatic layer*. It is of much greater thickness than the periostracum, and is built up of calcareous columns or prisms arranged at right-angles to the surface of the shell. The prismatic layer, just like the rest of the calcareous shell, consists of both inorganic and organic substances. The organic part forms a kind of framework upon which the calcium carbonate is deposited; the organic substance itself is known as conchyolin.

The prismatic layer also is secreted normally by the cells of the mantle edge only. It cannot increase in thickness after it has once been formed. The prisms of which it is built are often surrounded by very delicate sheaths of periostracal-like substance and each prism or column is transversely striped. In fact, each column looks like a little pile of microscopic coins. This prismatic layer often occurs in pearls.

The third layer to be referred to is the most important of all from our point of view. It is the innermost layer, forming the internal surface of the shell, except at the small areas where the adductor muscle is attached and where there is a still more internal layer of shell substance.

It is known as the *nacreous layer*, and consists of that valuable substance, mother-of-pearl.

The nacreous layer differs from the layers already

described in being secreted normally by the entire external surface of the mantle. Consequently, it can increase in thickness throughout the life of the animal. It is built up of very delicate lamellae or tiny platelets which are arranged overlapping one another (fig. 3), parallel to the surface of the shell. The edges of the lamellae are zigzag, and as the surface of the mother-of-pearl layer exhibits parts of the faces and also the edges of the lamellae, a surface is

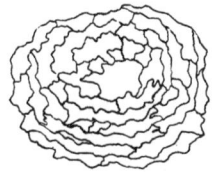

Fig. 3. Nacre or Mother-of-Pearl laminae in surface view (highly magnified).

formed which possesses by virtue of interference phenomena the so much desired iridescent lustre.

It is this nacreous layer which gives the pearls their beauty, and it is this layer which is often deposited over sand grains, or objects placed by man between the shell and mantle. Further reference to the nacreous layer will be made later.

The last layer to which reference will be made here is found between the muscle and shell. It is called the *Hypostracum*, and is, like the two preceding layers, formed of calcium carbonate. It is built up of distinct columns arranged at right-angles to the surface of the shell.

Now let us briefly look at the course of shell growth. The shell grows in size at the edge. Here the edge of the mantle is forming periostracum and prismatic layers, and there is practically nothing else. A little distance in from the edge of the shell the surface of the mantle is secreting nacreous or mother-of-pearl substance, and this goes on continually. Thus the shell increases in thickness over its whole inner surface.

The secretion of the shell is really very complicated, though the organs forming it may appear very simple. Shell formation is due entirely to the cells covering the mantle. These cells fabricate and pour out (or secrete) a mixture of organic and inorganic substances (conchyolin and lime). Crystallisation of these substances takes place altogether outside the animal and the shell is the product.

What determines the regularity of structure or the formation of beautifully marked or sculptured shells is unknown. All the cells seem practically alike, yet there must be some determining factor hidden there. If a mollusc shell is broken (or a piece of the shell is removed) some distance away from the edge, a most extraordinary thing happens

THE PEARL OYSTER

which is of the greatest importance in regard to theories of pearl formation.

The shell is repaired, but the repair substance is not, as was once thought, only nacre. The mantle cells, which normally would be secreting nacre, now secrete at the area of breakage, first periostracum, then prismatic layer, and finally mother-of-pearl.

Thus, although certain parts of the mantle surface secrete *normally* certain definite parts of the shell, the cells which secrete the nacre are capable of secreting all the other layers if necessary. This extraordinary property is made use of in the formation of pearls, as will be seen later.

Let us now leave the dead shelly covering and look at the animal itself which has been responsible for the secretion of the valves. The figure (fig. 4) is a view of the right side of the animal. In the centre there is to be seen the large adductor muscle, and this seems to take the place of a central scaffold. Dorsal to this, that is towards the hinge-line, is a soft unsegmented body in which lies the heart and the viscera. Ventral to the muscle are the flat plate-like gills hanging down as curtains into the cavity that is enclosed by the two valves. On the left the foot is to be seen with fibres of byssus. The mouth is situated above the byssus. This figure, with the rough description given above, represents what is seen when a flap of tissue (very important, as we shall see, from the point of view of pearls) has been removed. This flap, known as the *mantle* or pallium, is a lateral fold of the wall of the body of the animal. It is similar in shape to the valve of the shell; in fact it has formed the latter.

The cavity between the two mantle flaps into which one looks when a mollusc's shell is opened, is of course nothing else but part of the external world, and is known as the pallial cavity. The seawater enters this space and bathes the gills which hang down into it and also the internal faces of the mantle flaps. The mantle lobes are therefore separated anteriorly, ventrally and posteriorly, but continuous dorsally underneath the hinge-line. The free edge of the mantle is thickened and darker in colour than the rest. It also bears numerous short tentacles. In the common scallop the mantle edge bears numerous complex eyes as well as the tentacles.

It is usual to find the edge of the mantle thickened very considerably in the majority of bivalves, and, as we have seen, the edge is of great importance, for it secretes *two* of the layers of the shell, the external periostracum and the prismatic layer. These two layers are renewed by the mantle edge, and growth of the shell, so far as these layers are concerned, can only take place at the margin.

The inner nacreous layer is, on the other hand,

secreted by the whole face of the mantle in contact with the shell.

The mantle itself is formed of connective tissue with numerous blood spaces—it has an extensive blood supply. It is bounded by a layer of cells, the ectoderm, and the ectodermal cells, next to the shell, are secretory, as we have seen. The ectoderm facing the pallial cavity is ciliated, that is, the cells possess delicate vibratile processes or "cilia."

The *Foot* is an organ capable of very considerable expansion and contraction. It arises from the visceral mass just below the mouth.

The variation in size is brought about in the following manner. The foot is highly muscular, but it is also full of cavities which communicate with the blood system. If blood is forced into the spaces the foot increases in size and becomes several times as large as in the contracted condition. Contraction takes place by muscular action. It is by alternate expansion and contraction in this way that the common razor shell of our coasts bores its way down into the sand.

On the ventral surface of the foot and at the end nearest the body is a gland, situated in a pit. From this a groove runs to the end of the foot to another pit. The gland secretes the fibres known as byssus, easily seen in the common edible mussel, and in both cases the byssus is used for attaching the animal to the rocks or to other shells. All these structures are indicated in fig. 4. In addition, however, to the secretion of byssus and its attachment by the same, the foot is actually used for locomotion.

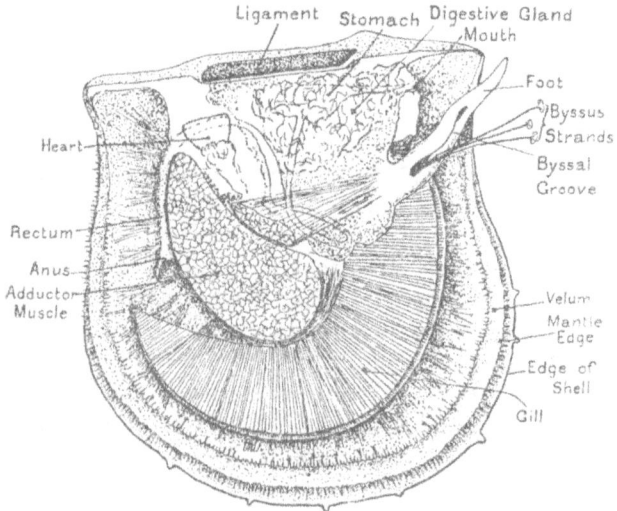

Fig. 4. Ceylon Pearl Oyster. Mantle and gills from one side removed. (After Herdman.)

The byssus can be pulled away, and then the pearl oyster, being free, extends the foot and uses it as a crawling organ in a somewhat similar manner to the snail. Having found perhaps a more suitable

THE PEARL OYSTER

position for re-attachment, the top of the foot is applied to some rock, and the pit at the end of the byssal groove is thus pressed against the attachment surface. The edges of the groove are now brought together, so that a narrow tube is formed, and down this tube passes a fluid secretion from the byssal gland.

It soon hardens in contact with sea-water, and after about five minutes the foot may be withdrawn from the attachment surface. It will then be seen that an elastic fibre extends from the mouth of the byssus gland pit to the point where the other pit was pressed against the rock. This process is repeated very many times, until a large number of fibres extend in a bunch from the byssus gland opening to the attachment surface.

If a pearl oyster or an edible mussel is pulled away with force from its attachment, the byssus is not usually detached from the foot. It comes away from the rock or it breaks. After some time, however, the mollusc sheds the old byssus, and if it re-attaches itself it does so by the secretion of a new one.

The Muscles.

Reference has already been made to the large central adductor muscle which pulls the two valves together and closes the shell. In many of our bivalves there are two adductor muscles. Where one only is present, it corresponds to the posterior of the two. This adductor muscle can contract with considerable force. The force required to open an edible oyster, an edible mussel or a cockle is probably known to most readers. This force is being expended in overcoming the strength of the adductor muscles. Whilst, however, the bivalves can resist very considerable pulls made for a short period, a very weak pull is sufficient, if sustained for some time, to break down the oyster's protection. This is the cause of much trouble, for one of the enemies of the bivalve, the starfish, opens the shell merely by pulling on both valves with comparatively little force, but with an insistence that is to be admired. The process is interesting, because for a long time it was supposed that the starfish poisoned the mollusc, when of course after death the shell would open automatically. This is not the case, as the starfish simply applies its arms to the two valves of the shell, attaches its numerous "tube feet" and steadily pulls. When its perseverance has been rewarded by the divarication of the two valves, the stomach is extruded and the animal proper of the mollusc digested.

In addition to the adductor muscle, there are several other large and definite muscles which are connected with the foot.

There are six of these altogether. All are

THE PEARL OYSTER

attached or inserted on to the shell valves. Two of them are known as Retractors. They originate in the walls of the byssus gland and pass backwards one to each shell valve. The other four are Levators, two of which are anterior and two posterior. By the contraction of these muscles the foot is moved from side to side or retracted.

The protrusion of the foot to the outside of the shell is produced by a rush of blood into the cavities of this organ.

There are numerous other smaller and diffuse muscles in the body which need not be detailed.

The Viscera.

There is no large cavity in the molluscs which, like the human body-cavity or coelom, apparently contains the viscera. The cavity which corresponds to this is small; it is termed the *pericardium*, and contains the heart. As in all bivalves there is a definite heart with blood vessels, but the vessels often expand into mere cavities in the tissues and it is difficult to speak of arteries and veins, except when referring to definite large vessels with cellular walls.

The heart, in the pericardium, is situated dorsal to the adductor muscle and between it and the so-called liver (see fig. 4).

The mouth of the animal is situated high up near the hinge-line, above the foot and consequently anterior in position. It is partially hidden by two lips which are produced at the sides to form flat plates—known as labial palps. Since each corner of the mouth is produced laterally to merge into the gutter between the two palps on each side (a feature not easily shown in a diagram), the two pair of palps act as directing walls and guide food to the mouth. The mouth leads into the alimentary canal, a more or less simple tube, which, with the exception of a swollen region, the stomach, is much the same in structure and appearance throughout its length.

It coils once and finally runs over the adductor muscle to open by the anus into the cavity between the two shell valves. The stomach is embedded in, and surrounded by, the so-called liver. It is very difficult, however, to compare any organ in Invertebrates with organs in Vertebrates, and this so-called liver possesses many functions that are not performed by the human liver. It is better to call it a digestive gland. It secretes juices that help to break up and alter the food and it *absorbs* this digested nutriment.

Now let us see how these organs do their work and by what means the pearl oyster obtains its food. Starting at the very beginning of the story, the food of the pearl oyster consists of minute animals and plants which occur everywhere floating in the seawater. These same microscopic organisms are found in the fresh water of ponds, streams, rivers and lakes

THE PEARL OYSTER

—and they occur in the sea, though of course different species are characteristic of all these different situations and are in fact characteristic and different for each season of the year.

These organisms must be brought within the shell, and as the bivalves are not capable of pursuing them, a current of water is caused to flow into the shell cavity which will bring them with it. This current of water is very important for another reason —the pearl oyster, like other living organisms, must breathe. It obtains its oxygen from the water, and consequently a current of water must enter the shell continually, bringing in oxygen, and must pass out again with carbonic acid gas, the product of respiration.

Thus an inflowing current of sea-water brings oxygen for respiration and food for nutrition at the same time. The water current is produced chiefly by the gills. We have already referred to these organs as being flat plates in the bivalve molluscs. They occur as two double plates on each side, hanging down like curtains from the adductor muscle and visceral mass, and occupying a considerable part of that cavity (itself a part of the external world) which is encompassed by the two valves of the shell. These are really only *two* gills, but as each one appears as two double plates (half gills) there are often said to be four. It must be remembered then that each one of the four is but a half gill. Two half gills belong to each side of the body.

Each gill plate is made up of a number of filaments, all hanging down side by side. They are held together by delicate hair-like processes, so that they appear to make up a firm sickle-shaped plate. As a matter of fact, a slight touch with the finger serves to separate the filaments so that a gill plate looks like a comb, the teeth of which are not brittle and strong, but soft, and only kept in position by adhering to one another. Perhaps a better comparison would be with the vane of a feather. Here also we have a number of delicate processes which are held together to form a flat plate.

The gills are covered with delicate microscopic hair-like processes or cilia, arranged in a regular and particular manner, and it is by the continual wafting of these cilia that the water current enters the shell. Probably all readers of this book have watched the effect of the wind, blowing across a field of corn. Successive waves appear to cross the field, but the motion of each cornstalk is to and fro. If one could make the cornstalks move in this way, of their own accord, the process would be reversed and a current of air would be produced. This is exactly what takes place on the gills: the cilia are the individual cornstalks—they move continually, but rhythmically one after the other, so that waves are

THE PEARL OYSTER

set up. This sets the water in motion. The current of water enters the shell between the two mantle lobes at a region about the middle of the ventral margin. It passes over the gill plates and through them. Passing through them continually, the water is to a certain extent robbed of its oxygen and leaves the shell at what is known as the exhalent opening, situated posteriorly. If however food particles enter with the water of the inflowing current, they are wafted along the gill margins anteriorly, until they reach the two flaps known as the palps. There are a pair of these on each side of the body; the outermost of each is really a prolongation of the upper lip of the mouth and the innermost a prolongation of the lower lip. Consequently the gutter or groove between either pair of palps leads to the mouth.

Like the gills, the palps are covered with cilia, hence the food particles are wafted along between them up to the mouth opening. The particles may be accepted as food (and then they pass into the alimentary canal) or they may be rejected. In the latter case they are picked up by a current of water running along parallel and close to the margin of the mantle and are carried back until the exhalent opening is reached. Here they leave the shell together with the current of water that has passed through the gills. In this manner the pearl oysters, and indeed also the cockle, mussel, and other bivalves obtain their food. They search not, neither do they place food in their mouths by special appendages.

Reference has been made once or twice to the gills. It must be said here that they are not the only organs concerned in respiration.

The mantle lobes themselves are very efficient respiratory organs, for they contain numerous blood spaces, and the wall separating this blood from the sea-water in the cavity of the shell is extremely thin and allows the interchange of the gases.

The Blood System.

All bivalves possess a heart and system of blood vessels. This perhaps is not common knowledge, because the blood of these molluscs is practically always colourless. The absence of the characteristic red haemoglobin of human blood does not render the blood different so far as function is concerned. There are other chemical compounds present instead, which help to carry out chemical changes similar to those performed by haemoglobin in human respiration.

The heart consists of one ventricle, a central, dorsally situated, muscular-walled bag, and two auricles, one on either side, with much thinner walls. The ventricle contracts or "beats" rhythmically, and at each contraction the contained blood is forced out through two arteries (not shown in fig. 4) to the

THE PEARL OYSTER

system. On dilatation after each contraction, blood is sucked into the ventricle from the auricles. The blood can only flow in one direction. It is prevented from re-entering the auricles, when the ventricle contracts, by valves which guard the orifices. The auricles receive their blood from the gills and mantle; thus the heart only contains blood of one kind—pure. The blood system is much more simple than that of man, where the heart has two ventricles, and pumps out both pure and impure blood.

The Excretory Organs.

The pearl oyster possesses two so-called "kidneys." They are simply tubes which open at one end into the pericardium and at the other into that part of the external world enclosed by the two shell valves.

Excrete matter is removed from the blood, which circulates in the walls of these tubes and is passed into the lumen, eventually reaching the exterior. Waste matter may also be excreted into the pericardium itself by the walls of the heart. This excrete matter also passes down the lumen of the kidneys to the exterior.

There now remains to be described the nervous system and sense organs, and finally the reproductive organs to render this general account of the anatomy complete.

The Nervous System and Sense Organs.

Since bivalves like the pearl oyster and the mussel do not make a very active search for food, but lead a sluggish existence, partaking of such nutriment as is wafted to them, they miss that active struggle for existence which is bound up with a search for food and a combat with enemies. Not that they take no part in a struggle for existence. That seems to be part of the ordained law of life for every organism. It is here, however, more of a passive nature. We are not surprised then to find that both nervous system and sense organs are but simple.

Here in fact is the explanation of the absence of a definite and obvious head with eyes and other organs. To the man in the street an animal lacks all individuality if it does not possess such a structure, on which he may interpret some expression. One cannot point to the animal's head, but merely to the head region. Specialised sense organs are few, and they cannot be compared to any possessed by well-known vertebrate animals. They are of use to the animal for orientating itself and for testing the water which continually streams into the shell. The edges of the mantle are sensitive to light, and a shadow thrown on to them causes an immediate closure of the shell. This is the sole means of defence possessed by the pearl oyster. As a matter of fact,

THE PEARL OYSTER

when eyes are present in the bivalves they are usually situated, not in the head region, *but on the mantle edges*, the only regions where they would be of any service. Thus do necessity and use triumph in nature over all else.

The nervous system consists simply of three pairs of small bodies or ganglia (which are aggregations of nerve cells) connected by nerve fibres. One pair, the cerebral ganglia, is situated in the head region at the sides of the oesophagus. Another pair, the pedal ganglia, is connected to the cerebral ganglia by two connectives, but the two ganglia are fused to form one mass at the base of the foot. The third pair is situated on the face of the adductor muscle. The ganglia of this pair are connected to the cerebral ganglia but not to the pedal.

Thus the nervous system is bilaterally symmetrical and composed of right and left halves.

In this simple invertebrate nervous system we find nerve fibres running out from these ganglia to muscles. Other fibres, sensory in function, run in from the peripheral parts of the body and the gills. A stimulus such as a difference in the constitution of the water, a shadow thrown on the mantle edge, or a blow struck by some external object on any soft part of the body, is followed by the transmission of nerve impulses to ganglion cells and from these to muscles. A response may or may not follow the stimulus, but we can only regard it as a reflex action.

The Reproductive System.

The last system in the body to which we shall allude is the one comprising those organs upon which the upkeep of the race depends. In the pearl oyster the sexes are separate, that is to say, an individual is either male or female. (In the scallop, both male and female organs are found in each individual.)

There is however no difference to the eye in the external appearance of male and female oysters; the sex can only be determined by examining closely the organs of reproduction inside the body. The organs themselves are rather diffuse, and consist of hundreds of branched tubules which cover the stomach, the digestive gland, and the intestine. The tubules open into little sacs, and in these the eggs or spermatozoa are formed, according to the sex. The tubules join up to form ducts which open into each other until they form a large duct on each side which opens to the exterior. When the reproductive organs are mature, the eggs or spermatozoa are shed into the sea. They pass out of the shell and float about in the water in which the animals are living. As the spermatozoa are developed in countless numbers and as the pearl oysters are gregarious, there is every chance of a sperm meeting an egg. Fusion of the two then takes place and fertilisation is accomplished.

Fertilisation depends then not upon any act of the pearl oysters themselves, except of course the emission of the eggs or sperms into the sea-water, but upon the accidental meeting in the sea of the two microscopic germs.

Where this method is adopted we always expect to find myriads of eggs developed in the individual, to make sure in the terrific mortality that afterwards occurs that some small but adequate proportion will survive. This is the case in the pearl oyster. It would be a tax on the imagination to picture what would happen if *every* egg produced by the pearl oysters developed into even a young oyster. This, however, applies to many thousands of other species in the animal world.

The description of the pearl oyster should give an idea of the type of animal that is popularly known as a bivalve. It would not be difficult for the reader to find most of the structures described in the edible oyster, the scallop, the clam, the edible mussel, cockle and fresh-water mussel. The three last-named might look rather different, but this is simply due to the fact that instead of possessing one large central muscle for closing the shell, they possess two, one at each end of the body.

CHAPTER IV

THE LIFE-HISTORY OF THE PEARL OYSTER, AND THE CONDITIONS UNDER WHICH IT LIVES

THE mature pearl oysters, as we have seen, discharge into the sea their sexual products, eggs or spermatozoa, according to the sex. The actual emission has been observed in May for Ceylon oysters, but probably the spawning season extends over many months of the year.

The eggs and spermatozoa pass out direct from the reproductive organs into the water, which is continually passing into and out of the pallial cavity. Fertilisation of the eggs takes place in the sea-water. There is however a very great probability, as the author has observed in the case of other lamellibranchs, that the emission of ripe spermatozoa from the male is stimulated by the presence of female oysters.

As the pearl oyster is a gregarious animal like our own edible mussel, there is every chance of the numerous eggs being fertilised, even though no act of copulation brings the sexual products together. This method of fertilisation of the eggs may be taken as holding good for the other relatives of the Ceylon Pearl Oyster—the Mother-of-Pearl Oysters—and also for our edible mussel. It does not hold good, however, for the fresh-water mussels of our streams and lakes, the life-history of which will be discussed further on. The small egg, at first irregular in shape, but becoming spherical after fertilisation, floats in the sea at the mercy of any current over the banks; but development continues, and we soon have a little multicellular embryo which swims by means of cilia. This power of locomotion is of course of no account in determining the wanderings of the embryo, which, during its entire life, is dependent upon the erratic currents in which it floats. The second day marks the appearance of a tiny shell which develops until a minute transparent bivalve shell encloses the larval oyster. It is still a swimming form living as a floating organism. This brings us to the stage immediately before the formation of "spat." The term "spat" is given to the earliest attached stage in the life-history of molluscs like the pearl oyster. The attached stage is probably reached just within a week after fertilisation has taken place.

It will be seen that the chances are very great that during this time the free swimming embryo has been carried by tidal currents far away from the home of its parent oysters. It may perhaps be taken over a great oceanic abyss, or a sea-floor quite unsuitable for its attachment, in which case the adult stage would never be reached. It is often quite easy to provide artificial supports to which the pearl oyster and other shellfish spat can attach themselves when natural objects are wanting. Thus, for example, at places on the French coasts, attachment areas are prepared for the edible oyster and mussel. Drain pipes, placed in shallow water, have served as very favourable objects for the young spat of the scallop.

The shell valves now continue to grow, increasing in thickness, but retaining for some time their transparency. The larval swimming organ becomes reduced and disappears, and the shell-closing muscles and foot develop. These tiny oyster spat are only about 0.1 mm. in diameter.

If they have only sandy wastes as sea bottom it is certain that their life is short. Hence there is immense need for favourable grounds, with either hard objects or with seaweeds to the fronds of which the young spat can attach themselves.

The young pearl oyster is an actively moving organism, even though it attaches itself by means of a byssus. In this again it resembles many other common bivalves. The method of locomotion can be well seen if a quite small active edible mussel or a very young scallop is placed in a glass vessel of seawater.

The animal attaches itself to the glass by means of strands of byssus secreted in the way already

IV] LIFE-HISTORY AND ENVIRONMENT 45

described. Holding on by these strands the animal pulls itself nearer to the point of attachment, then stretches out its foot and proceeds to secrete other byssal threads and to attach them further away, on the glass sides of the vessel. As soon as a firm hold is obtained the old byssal threads are thrown off by the animal and remain adhering to the glass. This process is repeated time after time, and the track followed by the mollusc is indicated by the series of cast-off byssal threads.

Apparently the young pearl oysters do not travel so fast as young scallops (which as a matter of fact are famous throughout life for their powers of locomotion). They are sensitive to light and, in aquaria, move but little during the day. Probably this does not apply at the bottom of the sea on the pearl banks. During the first two years of life the pearl oyster grows rather quickly. The age and the size at which they are fishable, vary for the different species. The Ceylon pearl oyster has to battle with its enemies until it is four and a half to five years old. It is then roughly about three inches in height and two and a half inches in breadth.

The life-history of the fresh-water mussels in rivers and lakes is very different. The mussels live in clear water and on stony or sandy ground, either isolated or in great numbers.

They do not emit eggs outside their shell, for

these and also the young larvae which might develop would be swept down by the current into the sea. The eggs are kept inside the parent and develop into peculiar larvae with tiny shell valves provided with hooks.

These young mussels are set free and clap their shells vigorously. If by a happy chance they strike some fish swimming in the water, they attach themselves and actually live partially embedded in the tissues of the fish for some weeks. At the end of that time they drop out as young mussels and take up their normal life on the bottom.

It will be advisable now to look more closely at the conditions under which some of these pearl-producing molluscs live.

As it is beyond the scope of this little book to discuss the homes of all the various pearl and nacre-producing molluscs, it will be more satisfactory to glance at the habitat of the Ceylon pearl oyster as an example.

The animals are found mainly on certain banks, which are part of a submerged plateau in the Gulf of Manaar, off the north-west coast of Ceylon. These banks are known as "paars"—a name which implies a hard bottom of any kind. They are situated in depths of from five to about ten fathoms.

The pearl oysters, like most of our molluscs, require a particular kind of sea bottom. In addition

IV] LIFE-HISTORY AND ENVIRONMENT 47

to suitable conditions of depth, temperature and water, they must have some objects to which they can firmly attach themselves. One does not find pearl oysters, therefore, on the fine ooze which exists further out from the island at greater depths. Again, pearl oysters are not found on the hard bottoms from twenty fathoms down to a hundred fathoms, although in many of these places large encrusting organisms like corals and sponges are found in great abundance. The pearl-bank plateau off Ceylon is broken up into a considerable number of paars, from three to eighteen miles from land. The plateau is for the most part composed of sand, but here and there the actual bare rock appears, although in many places it is only covered by a few inches of shifting sand. This sand was examined by the author in connection with Professor Herdman's visit to the pearl banks. It consists of, sometimes, 40—50 $°/_o$ of skeletons of certain protozoa called Foraminifera, many of them attaining the size of a threepenny-piece. The remainder is made up of sand grains and calcareous animal remains. The rock itself is only a "cemented sand," for the sand grains in this sea-water with the calcareous animal and plant remains become cemented together. Lumps of coral and encrusting animal skeletons play their part, and we have in consequence a modern rock formation before our eyes.

Many of the rocky blocks—composite structures

of the inorganic and organic worlds—are veritable "marine biological stations" in themselves. Formed largely of the remains of corals, foraminifera, sponges, calcareous algae and sand, they are burrowed into and provide homes for tiny crabs and shy crustacea. Worm tubes, in which live beautiful tube-worms, wind about the surface. A patch of a colonial polyzoan, allied to those incrustations often found on seaweeds round our coasts, and perhaps a dozen or so pearl oysters hanging on by the byssus may complete the catalogue of some of the more conspicuous organisms.

That arch enemy of the mollusc all the world over, the starfish, is certain to be found creeping about amongst its fare, and sea urchins and sea cucumbers (close relatives of the starfish and not vegetables, as the name might imply) represent the Echinodermata.

The well-known seaweed-covered rocky beds of our own coasts are very different. The seaweeds in these tropical waters are enclosed in lime and might well be mistaken for non-living things. In many cases the plants live enclosed in the bodies of the animals! Indeed, in these tropical regions where the nitrogenous compounds so essential to plant life seem to be deficient, the vegetable cells have been driven to live inside the animals. This phenomenon —of plant animals—is represented of course in our

IV] LIFE-HISTORY AND ENVIRONMENT 49

climes by the green hydra, and that interesting worm so ably described by Prof. Keeble in one of the volumes of this series.

The basis of a paar may be said to be those hard bodies to which young and old oysters can attach themselves. This is known as "culch," and consists of dead corals, lumps of calcareous algae, and old shells of various species. In many places there are suitable banks where the area occupied by sand alone would be available for pearl oyster occupation if "culch" were present.

Artificial culching, that is, the bringing of suitable material, such as shells, coral blocks or rubble from other places and depositing it on the sandy bottoms, might easily increase the extent of the oyster beds. Somewhat similar methods have been adopted on the coasts of France and Holland, and various structures have been built up for the attachment of young edible oysters or mussels.

If we consider all the surroundings, solid and liquid, living, dead and inorganic, as making up the environment of the pearl oyster, we find there are favourable and detrimental features. The favourable conditions have been touched upon; let us now look at the factors which lead to destruction. One of the most important agents in causing the death of the pearl oysters is the shifting of sand.

This is occasioned by strong currents raised by

monsoons and storms. The oysters are buried—young oysters are destroyed, and in some cases the destruction is enormous. Herdman and Hornell observed the loss of "thousands of millions" of young oysters from this cause in a year. On some of the paars this depletion seems to be of regular occurrence every year.

Next we come to the natural enemies of the mollusc, and they are numerous, though probably only a few commit serious damage. The "black-listed" animals are in the first place the fishes which devour oysters. They are chiefly skates and rays of different species, and the file-fishes. The large rays, up to a dozen or more feet across, have been observed feeding upon the shellfish. With their broad-surfaced, specially adapted crushing teeth they can make short work of the delicate shells; and shell fragments, the result of their depredations, are often numerous. The file-fishes feed upon immature oysters.

The next group of enemies is that of the molluscs which bore into shells. The gastropoda (univalves like the common whelk of the English coast) are provided with a proboscis armed with a chitinous file. In addition, however, to boring holes by this means and then inserting the proboscis, it is quite certain that many of these univalves either catch an unsuspecting bivalve with its shell open, or drive part of their shell between the valves of their prey and

IV] LIFE-HISTORY AND ENVIRONMENT 51

then attack the muscle. If this happens, the empty shells of the oyster alone are left, perhaps uninjured and with nothing to indicate the mode of death.

Certain sponges (*Clione*) bore into the thickness of the shell valves just as they do in the scallops of our seas. The valves are honeycombed in all directions until they become completely rotten. More serious, however, than these animals are the starfish. Herdman records the destruction of between 200 and 300 a day when the s.s. *Violet*, in the fishery of 1905, was dredging for pearl oysters. This must give a huge total for the starfish living on the paars.

Starfishes have, too, such powers of regeneration that, other conditions being favourable, they are little disturbed by being torn into two. They are very voracious, and have reduced the opening of oysters to a fine art! In one case Herdman and Hornell considered that the disappearance of $5\frac{3}{4}$ million oysters in a year was probably due to this enemy.

So much for the environment. It must not be forgotten, however, that to any one oyster the other brother oysters are also parts of the environment. If therefore the young oysters are overcrowded, a large number will die through interference with nutrition. The competition for food is a greater factor in the history of this universe than perhaps many of its inhabitants believe!

The difficulties under which the pearl oyster lives have been briefly touched upon. Finally, man himself, by his heedless removal of millions of breeding oysters, is not the least of the enemies of the pearl oyster, and a failure of the pearl-fishery must often be attributed to his influence.

CHAPTER V

PEARLS

THERE is no little difficulty in making out from the literature and from "common knowledge" what is actually meant by the term "pearl." Some dictionaries are most ambiguous on this point. For example, in the same paragraph we are told that a pearl is "a silvery white smooth and iridescent gem, extracted from the pearl oyster"; "something round and clear, like a dewdrop"; "anything very precious." No doubt the term pearl is applied to many beautiful objects in poetry and literature, but so far as we are concerned, the name will have to be used in a more restricted sense.

A pearl consists of a number of layers of organic and inorganic matter, arranged more or less regularly round some common centre (termed the nucleus) and resembling in composition and microscopic structure certain layers of the shell of the mollusc in which the pearl has been formed. It may be of almost any colour and may lack altogether the lustre and general appearance that is commonly associated with pearls. As a matter of fact pearls are nothing more than abnormal products, calcareous pathological bodies formed, owing to some irregularity, in the tissues of certain shellfish.

Von Hessling gave the following as a definition of pearls. "Pearls are shells converted into a spherical form; they possess the same histological, physical, and chemical peculiarities, in so far as these do not belong specially to the rounded shape, and both undergo, in all the different stages of formation, the same fate."

As we have already, for this special purpose, described the shell of a bivalve in the account of the pearl oyster, it will be easy to apply this knowledge to the structure of pearls.

It is necessary, however, before doing this to draw attention to other calcareous bodies which are found in molluscs, which may possess a pearly lustre, and yet are not true pearls. These bodies are known as *blisters*. They are very often formed owing to foreign objects such as sand grains getting between the shell and the mantle; they often arise in the process of repair of shell perforations.

"Blisters" have a different mode of origin, therefore, and they are usually attached to the shell. They may be very large, and are often cut away from the shell and used in cheap jewellery. It is blister formation that has been encouraged, as we shall see in a later part of this chapter, by artificial processes for trade purposes.

True pearls themselves are of two kinds,

(a) Cyst pearls (Herdman) or Parenchyma pearls (Jameson),

(b) Muscle pearls (Herdman).

Of these, the former are the most important commercially, and have formed the subject of most of the scientific work on the origin of pearls. They are usually spherical pearls which occur singly in the mantle or in the soft tissues of the pearl oyster or mussel. They may occur even some distance away from the surface, embedded in the digestive gland or so-called liver.

Muscle pearls were distinguished first by Herdman (British Association Report, Southport Meeting, 1903). They occur usually in large numbers in the muscular tissue, where the muscles are attached to the shell. Fig. 5 indicates the positions where these pearls are found in the pearl oyster of Ceylon.

Now, putting the definition of von Hessling in another form, it may be said that pearls are bodies consisting of layers of calcareous material with an organic basis. The layers are arranged in a more or less concentric manner, round a central nucleus, and the four types of structure met with in the shell may occur in the pearls.

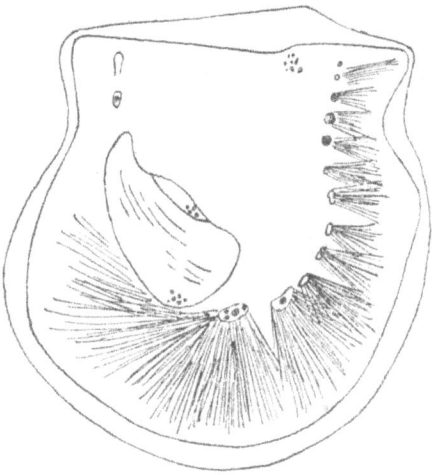

Fig. 5. Diagram showing the positions most frequently occupied by muscle pearls in the Ceylon Pearl Oyster. (After Herdman.)

These layers, it will be remembered, are the periostracum (the most external, almost entirely organic layer), the prismatic layer, the nacreous layer (or mother-of-pearl substance), and the hypostracum (the layer found where the large closing muscles are attached to the shell). Some pearls consist entirely of concentric laminae of a substance like the periostracum. These pearls are not common but are sometimes to be found near the very edge of the mantle. Naturally, they are somewhat brown in colour and lack entirely the characters which are associated with pearls in the popular mind.

Pearls consisting entirely of hypostracum-like substance are to be found, often in large numbers, in the muscles. In fact, many if not most of the muscle pearls are composed of hypostracum.

Now, where the muscle pearls are abundant, large numbers of calcareous concretions or calcospherules are scattered in the tissues. Herdman and Hornell suggested that the muscle pearls were formed around these microscopic calcospherules. It is extremely probable, as suggested by Jameson, that these calcospherules are in reality small pearls—probably of a hypostracum-like substance. Round them may be deposited more layers of hypostracum or nacre. There is no evidence in any case to support the view of Mr Southwell that the calcospherules are calcareous depositions from the blood.

The pearls composed of hypostracum may possess either an empty central cavity, a cavity containing granules, or a small central nucleus of yellow periostracal-like substance. The calcium carbonate is laid down round the centre in a manner strikingly

PEARLS

like the hypostracal layer of the shell. Radiating striae pass outwards, and in addition a delicate concentric lamination can be observed.

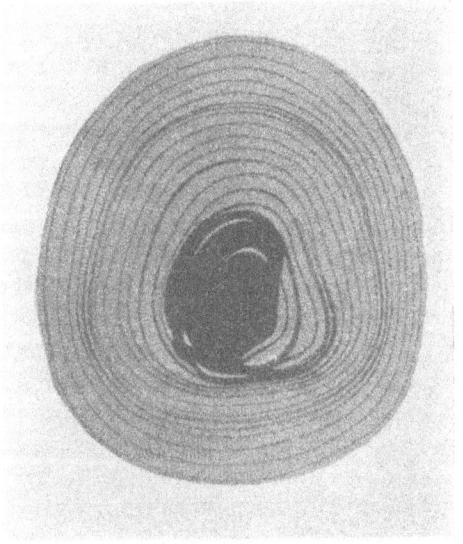

Fig. 6. Section of a pearl composed of prismatic substance with nucleus of periostracum-like substance. (From Rubbel.)

Pearls of larger size, consisting almost entirely of prismatic-like shell substance are more common, but it must be remembered that usually the periostracal substance occurs too, both in the very centre as a nucleus, and also in concentric layers alternating with the prismatic substance (fig. 6).

The prismatic substance resembles closely that same layer in the shell, in the manner in which it is deposited, but the columns or prisms are radially arranged in the pearl so that they are at right-angles to the concentric laminae.

Most pearls consist of a nucleus, to which reference will be made later, and more than one, perhaps all, of the different varieties of shell substance. The most important of these is of course the nacre.

The figure below is that of a nacreous pearl composed of two pearls which have fused together. The nuclei consist of periostracal substance and also hypostracal substance. Round these nuclei there are concentric laminae of nacre with two rings of periostracum. These concentric layers of periostracum may spoil the appearance of a pearl, composed otherwise almost entirely of nacre. In the same way, layers of prismatic substance may interfere with the continuous deposition of the nacre.

Owing to the deposition of the layers in concentric form, it is possible for a clever jeweller to strip off the laminae, just as the "skins" can be removed from an onion, though of course the laminae of nacre are never continuous right round a pearl.

This process of skinning the pearl consists in

removing the outer layers, if they are bad. There is a great element of chance in the process, because one never knows whether the deeper layers will be better or worse, and each layer removed reduces the size and value of the pearl.

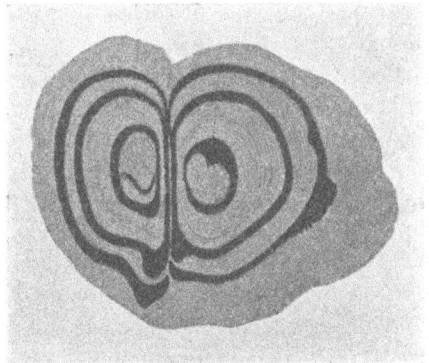

Fig. 7. Section of nacreous pearl, really composed of two fused pearls. Note the nuclei of Hypostracum and the concentric dark layers of Periostracum. (From Rubbel.)

In any case the external appearance of a pearl gives no clue to its internal structure, and the reason for this follows from the description of the structure given above.

There is some little difference between the nuclei of pearls. Thus the pearls from the common edible mussel (Mytilus) have often very large nuclei. The pearls from the Ceylon pearl oyster on the other hand have small nuclei. The nucleus is an extremely important body because of the part played by it in the origin of pearls.

It is, moreover, around the question of the nucleus that the various theories of pearl formation have been built up.

The nucleus, in fact, is supposed by its presence to have stimulated the shellfish, so that it secreted the various layers of shell-like substance around it and thus formed a pearl; or, more correctly, to have caused the formation of a little sac of cells which poured out these same substances. The nucleus may be a foreign body or it may actually be one of the substances found in the shell layers. We have seen for example that in many pearls the nucleus is a substance like the periostracum or the hypostracum of the shell.

In other cases, however, the nucleus is a microscopic animal parasite, a young larval worm which has penetrated into the body of the shellfish and died because of its encasement in a tomb of pearl. Sand grains and other hard fragments from the external world may also be found as the nuclei of pearls.

The composition of pearls must vary according to whether they consist of periostracum or of the other more calcareous layers. Leaving out of the question the pearls of periostracum and taking pearls possessing an outer covering of nacre, we have the following analyses:

Harley.

Calcium carbonate	91·72 %
Organic matter	5·94
Water	2·23
Loss	0·11

Dubois.

Calcium carbonate	91·59 %
Organic matter	3·83
Water	3·97
Residue	0·81

An analysis of nacre from the Ceylon pearl oyster shells gave the following (Roaf):

Calcium carbonate	88·79 %
Calcium sulphate	4·93
Organic matter	2·32
Water	2·28
Loss	1·68

The calcium carbonate occurs in the form known as aragonite.

So far as hardness is concerned it may be said that some pearls require a hammer to break them. This property varies greatly with the structure of the pearl.

Moebius found that pearls were scratched by

apatite: in Mohr's scale, this occupies the position indicated in the following:

Apatite 5, Orthoclase 6, Quartz 7,
Topaz 8, Diamond 10,

and the pearl is therefore easily scratched by many other stones. The specific gravity of oriental pearls varies, but the average may be taken as being from 2·650 to 2·686. The specific gravity of fresh-water pearls seems to be higher, *i.e.* 2·724. Poor pearls are not nearly so heavy and the specific gravity may be much less—averaging 1·5 in fact.

In addition to the smaller spherical true pearls, there occur irregularly rounded bodies, often of very considerable size, known as baroque or baroche pearls. These are sometimes mounted in jewellery so as to hide their irregularity, but many have been used so that their very irregularity added to their value. At the beginning of the 18th century there was quite a popular craze for baroche figures. Curious, quaint men and animal figures were constructed with the aid of the pearls and precious metals. Many of these are to be seen in the British Museum. There are good collections, too, in the Louvre and in the Dresden Museum.

At the present time small baroche pearls are being mounted in necklets and bracelets.

The colour of pearls varies very considerably, and all the possible shades between pure white and black are known to exist. White, or as nearly white as is possible, is usually regarded as the most sought after and perfect tint. Colour alone, however, is not nearly sufficient and there must be in addition a peculiar kind of lustre.

It is this combination of tint and lustre that has made the Ceylon pearl of such great fame and value, and although equally good specimens often occur in other places, even in fresh-water molluscs, on the average the Ceylon pearls seem to be better than those from most other localities. Many of the colour effects are not due to the presence of actual pigment but rather to reflection and refraction phenomena. Thus one may have white pearls with a blue, a green, or a pink sheen, although yellow is by far the most common. Other pearls with more definite colours and darker tints generally possess little or no lustre, and some of these would hardly be considered true pearls at all by dealers.

This often applies to the pearls found in the common edible mussel. In fact, supporting this remark, we find in the glossary of one book on pearls a definition of true pearls as follows: "Pearls formed of nacre as distinguished from similar formations which are not nacreous." It is not at all necessary for a true pearl to have any of the nacreous layer represented in its structure.

The black pearl does not seem necessarily to be

black! A really black pearl is a gem of some value, but most so-called black pearls are merely deep shades of brown, or dirty blue. Nearly all the varieties of colour are to be found amongst freshwater pearls. This agrees with the statement made long ago by the Venerable Bede to the effect that the British pearls were "of all the best colours—that is, both red and purple, jacynth and green, but principally white." It is quite easy to see that a series of selected pearls in a necklace will be of greater value than individual pearls of the same size. It is said in fact that if you match a pearl with another, you double the value of each of them. The reason of course is the difficulty experienced in finding amidst specimens which vary so much in tint, lustre and shape, two that will stand the test of being viewed together. For definitions of some of the many names applied to pearls, reference should be made to the glossary at the end of the book.

It is not surprising, in the present day of imitation, that we should find remarkably good artificial pearls. By artificial pearls I mean actually manufactured articles, and not those structures which are secreted by the molluscs after stimulation by man. That is artificial stimulation, and the "pearls" produced are eally only "blisters." Very many of these imitation pearls come from France, where, as a matter of fact, the industry was created.

Globules of glass are used in one method. These are made of an opal-like tint and then treated with a special mixture known as "essence d'orient." It "consists of the iridescent guanin from the scales of a fish—whitebait." The scales are removed from the fish, and rubbed in fresh water until they lose the iridescent substance. This is abstracted from the water and kept in ammonia. A drop of the essence is introduced into the *inside* of the bead of glass and the internal surface of the latter coated uniformly with the mixture. The bead is then filled with wax. Other variations of this method are used.

Imitation pearls are sometimes made by cutting out spherical pieces from mother-of-pearl shell and polishing them.

Even artificial mother-of-pearl can be made now in pieces of considerable size. Substances allied to celluloid are prepared which are treated with "essence d'orient." Naturally the structure of these imitations whatever be their appearance is quite different from real mother-of-pearl.

The trade in these purely artificial products is of considerable value and imitation pearl necklaces are to be seen in shops in some of the best London streets.

We cannot close this chapter without referring in conclusion to the attempts that have been made to stimulate shellfish to produce true pearls. This is one of the lines along which biological science has at various times exerted its influence. As a matter of fact, the artificial production of pearls has so far proved a greater commercial success than the cultivation of the pearl oysters. Great care must be taken, however, to distinguish between true pearls and blisters, and most, if not all, of the artificially stimulated pearls are really only blisters, and so of comparatively little value.

The natives of the far East recognised many years ago that if a foreign body was insinuated between the shell and the mantle of nacre-producing molluscs, the irritating object would be covered with a layer of nacre. Sometimes small worms or even fish get into this strange position, and become coated with nacre (fig. 8). A beautiful specimen of this is to be seen in the British Museum. Hague describes how the Chinese carried on a manufacture of pearls in this way in the fresh-water bivalve *Dipsas plicatus*. The shellfish were collected and kept living in lakes for this purpose. They were first opened (usually by children) and then objects of various kinds (grains of sand, pellets of nacre, or images of Buddha) were inserted between shell and mantle. After about ten months or longer, up to three years, the shellfish were fished up and the nacre-covered objects removed and sold.

The nacre-covered pellets were naturally the most

like true pearls. This industry of the Chinese has been of considerable value and of some importance in the districts where it was carried on. The practice has, moreover, been rather general in the East and it is applied now on a larger scale by the Japanese. There is no reason why the method should not be successful in any mollusc, the shell of which possesses a pearly nacreous layer. The pearls produced are only blisters and attached to the shell, but in these days of cheap jewellery, a little care in mounting hides the weak spots and blemishes.

The great Swedish naturalist Linnaeus turned his attention to the production of pearls by artificially stimulating shellfish, and with a reference to his work we shall conclude this chapter.

In various works on pearl formation, references occur to the "secret process" of Linnaeus, but what this may have been was not known until quite recently. That Linnaeus *did* produce pearls seemed certain, and most writers on pearls simply made guesses as to the *modus operandi*. Herdman's curiosity was aroused by the numerous references, and by a search through the manuscripts in the Collection of the Linnean Society, London, he arrived at the truth. The letters and other documents are extremely interesting and begin with a letter to Colonel Baron Funck (dated Feb. 1761), in which Linnaeus states "that he 'possessed the Art' of impregnating mussels, and that he offered to make known the secret for the public benefit and use on condition that the state would give him a suitable reward[1]." The secret method seems to have been the following:

1st. A small hole was made in the mollusc shell;

2nd. A fine silver wire was inserted bearing at its end a small rough fragment of limestone;

3rd. The artificial nucleus had to be kept near the ends of the shell so as not to irritate too much the animal's body;

4th. The nucleus was kept away from the shell by means of the silver wire, otherwise the pearls would have been connected to the shell by nacre deposits.

The mollusc used in the experiments of Linnaeus was the fresh-water mussel.

CHAPTER VI

A CEYLON PEARL FISHERY.

An account of a Ceylon pearl fishery will best serve as an example of the oriental industry. The pearl fishery of this part of the world is probably of very great antiquity, and it was very likely carried

[1] Herdman, Presidential Address to the Linnean Soc., May 1905 (see *Proc. L. S.*).

on 2000 years ago in much the same simple way as at present. We have already seen how the Singhalese records go back hundreds of years before the Christian era. Let us now look at one of the recent fisheries. The Ceylon fishery is and always has been an intermittent one. At the present date it does not exist at all. Records show that throughout the centuries there has been the same sequence of productive years and bad years. More recently the hand of science has been called upon to alter the stern decrees of nature, but the call seems to have come too late to stave off the present cessation of the industry. In the 19th century there were only 36 fisheries. Sometimes, after a few good years there would be a long spell without any fishery.

How long the present unproductive period will last, none can tell. Neither can one say yet what combination of circumstances will cause the replenishment of the submerged banks with pearl oysters, although we may guess at the causes of depletion.

Herdman's visit to Ceylon in 1902 was after twelve years of unproductiveness, and it was just at a time when the banks were beginning to show signs of another fishery. The promise was fulfilled and for some years there was a truly remarkable series of fisheries—fisheries of very great value to the controllers. In fact, as more than one writer has said, the pearl fishery is the greatest gamble of the world. From the very commencement to the end the procedure is guarded and cared for by "fickle fortune," and it requires very little knowledge of the East to appreciate the excited condition of the natives of Ceylon and even India and more remote parts during a fishery. In olden times it was rumour which collected the divers. They heard from afar that the harvest was ripe and they came to the banks with zest to take part in the lottery. For some hundred years, however, the Gulf of Manaar banks have been carefully examined every year by a Government inspector, and the Ceylon Government has decided whether there would be a fishery or not. The fishery takes place in February, March or April, while the inspection was usually in November.

The fact that a fishery is to be held is now made known by the newspapers, but probably passes just as quickly in the more ancient way from mouth to mouth until the whole of the Orient is informed of the fact. The advertisements in the newspapers are printed in Cingalese, Tamil and English. A notice of the fishery of 1905 is appended.

A CEYLON PEARL FISHERY

"*Ceylon Pearl Fishery of* 1905.

"Notice is hereby given that a Pearl Fishery will take place at Marichchikkaddi in the Island of Ceylon, on or about February 20, 1905.

"The Banks to be fished are the South-West Cheval Paar, which is estimated to contain 3,500,000 oysters, sufficient to employ 200 boats for two days with average loads of 10,000 each a day; the Mid-East Cheval Paar, estimated to contain 13,750,000 oysters, sufficient to employ 200 boats seven days with average loads of 10,000 each a day. The North and South Moderagam, with 25,700,000 oysters, sufficient to employ 200 boats for thirteen days with average loads of 10,000 each day; the South Cheval Paar, estimated to contain 40,220,000 oysters, sufficient to employ 200 boats for twenty days with average loads of 10,000 each a day; each boat being fully manned by divers.

"2. It is notified that fishing will begin on the first favourable day after February 19. Boat owners and divers should be at Marichchikkaddi by February 16.

"3. Marichchikkaddi is on the mainland, eight miles by sea south of Sillavaturai, and supplies of good water and provisions can be obtained there.

"4. The fishery will be conducted on account of

Government, and the oysters put up to sale in such lots as may be deemed expedient.

"5. The arrangements of the fishery will be the same as have been usual on similar occasions. Persons attending the Fishery Camp from India will be permitted to travel to Ceylon by either of the following routes: (1) Tuticorin to Colombo or (2) Paumben to Marichchikkaddi—and by no other. Arrangements will be made as at the last fishery for travellers to proceed from Paumben direct to the Camp. The only restriction imposed on travellers by the Paumben route will be inspection by the Medical Officer at Paumben.

"6. Drafts on the Banks in Colombo or bills on the Agents of the Government in India, at ten days' sight, will be taken, on letters of credit being produced to warrant the drawing of such drafts or bills.

"7. For the convenience of purchasers, the Treasurer at Colombo and the different Government Agents of Provinces will be authorised to receive cash deposits from parties intending to become purchasers, and receipts of these officers will be taken in payment of any sums due on account of the fishery.

"8. No deposit will be received for a less sum than Rs. 250.

"9. All communications regarding the fishery must be addressed to the Government Agent,

A CEYLON PEARL FISHERY

Northern Province, Jaffna, Ceylon, up to the end of January, after which date they should be addressed to him at Marichchikkaddi.

"By His Excellency's Command,

"A. M. ASHMORE,

"Colonial Secretary.

"Colonial Secretary's Office,
 "Colombo, *December* 16, 1904."

This fishery of 1905 was of very considerable value. The temporary town of Marichchikkaddi was larger than ever known before. The inhabitants crowded in to it and were at least 10,000 more than in the previous record fishery of 1904. The numbers of oysters fished and the money obtained for the Government of Ceylon were far beyond all previous records. In fact, 81,580,716 oysters were taken and the revenue alone was £167,381. In the three years 1903, '04 and '05 the revenue amounted to £293,735, quite enough to show the importance of this fishery to Ceylon.

Now we may proceed to describe, first the events of inspection which lead to a fishery, and secondly the fishery itself.

There are numerous preparations necessary before sailing for the pearl banks. Boats have to be repaired and painted, old stores overhauled and new ones procured, and at the same time supervision given to a steamer which may serve as an oyster-dredger. The place selected as the centre of the pearl fishery in 1905 was, as stated above, Marichchikkaddi, near the mouth of the Modragam River. It is necessary to make an inspection of the beds of oysters which it is proposed to fish immediately before a fishery opens, so that the inspector, or commander-in-chief of the fisheries, as we may term him, may be in a position to mark off the ground and regulate the number of boats and days allotted to particular areas. The essential features of the preparations are as follows: Three flag-buoys are laid out by attendant launches in the direction of each cardinal point of the compass, at distances apart of $\frac{1}{4}$-mile, the inmost buoys taking their distance from the inspection vessel, which is anchored to serve as a pivot mark in the centre of the circular area to be inspected. Then four boats (usually modified whale-boats), each manned by a crew of six, together with three divers and two munducks, under the charge of an experienced coxswain, take up equidistant positions between the ship and the first flag-buoy and row slowly round the ship, retaining with wonderful accuracy their relative positions.

At regular intervals the crew rest on their oars to allow the divers to make descents and bring up oysters if any are present. The result of each dive is reported to the coxswain of the boat, who records the condition of bottom and oysters upon a diagram form with which he is provided. The oysters are retained in the boat for the inspector to examine. The four boats, having each made a complete circuit, are next ranged in line abreast in the same manner as before, between the $\frac{1}{4}$-mile and the $\frac{1}{2}$-mile buoys, and each then makes a second circuit. Lastly, there is a third series of circles, so that the four boats thus make a total of twelve concentric circuits, each boat making three. The results shown upon the four coxswains' diagrams are transferred by the inspector to a final diagram or plan furnished with twelve concentric circles. When this is done the inspector possesses a graphic diagram of the average distribution of old and young oysters, and the places where no oysters occur at all. He calculates in square yards the area occupied by oysters, and then the approximate number of oysters thereon may be estimated by taking the average number of oysters per dive (ascertained by examining the divers' results), in conjunction with the average amount of ground which a diver is credited with being able to clear at one descent. Usually this area is considered to be about three square yards, and by assuming this to be a maximum area the danger of overestimating the number of oysters is avoided. Inspection estimates are usually less than the total number of oysters obtained at the ensuing fishery.

The actual number of oysters is, however, not the main object of the fishery, the important question being the determination of the probable value of the pearls which will be obtained. It is necessary, therefore, for an official valuation to be made of the gems obtained at the inspection for the purpose of advertising the fishery. Consequently, during intervals, three large samples of oysters of fishable age are obtained, partly by means of divers and partly by dredging. The weight and the number of the pearls to each 1000 oysters are calculated, the particular place from which the oysters are taken being noted.

The next procedure is to mark out the fishing ground, and to make known the meaning of the marks. The method is rather a novel one. The boundaries of the fishing ground are marked by buoys bearing red flags, while a series of white-flag buoys are placed wherever fishable oysters are present. The divers are instructed to cluster their boats round the various white flags. The oyster buoys are placed according to the distribution of oysters mapped out after the inspection, but to guard against the possibility of mistake the abundance of oysters may be verified by preliminary dives before anchoring the buoys in position. The flag system may be further improved by marking the flags individually with distinctive numbers and signs.

Advertisements are then published throughout the East, especially in the vernacular, in papers reaching the Persian Gulf and the two coasts of Southern India, at the instance of the Colonial Secretary's office at Colombo. As a result of these advertisements divers, gem-buyers, speculators, moneylenders, petty merchants and persons of very diverse occupations, make speedy arrangements for attending the fishery. Indian and Cingalese coolies flock by the thousand to the coast, longing to play even humble *rôles* in the great game of chance. The "tindals" and divers provide boats and all essential gear for the work afloat, while ashore the Government supplies buildings and various forms of labour. Stories of the mushroom growth of towns wherever gold is found, or diamonds discovered, or oil struck, have become quite commonplace. Tales of the uprising of Klondike, Coolgardie, and South African cities fade beside Marichchikkaddi—the city with no foundation. Among its thousands of inhabitants are only a few hundred women who merit the right of being present through serving as water-carriers to camp and fishing fleet. This place, with its unpronounceable name, is the pearl metropolis of the universe. Probably there is not a stocked jewel case that does not contain gems that have filtered

through this unique city by the sea. It is a place
that comes and goes like the tide's ebbing and flow-
ing. A sand-drifted waste lying between the jungle
of the hinterland and the ocean is transformed by
the "open Sesame" of a fishery proclamation into
a seething mass of working humanity in a few
weeks. Sheltering roofs are erected, and a struggle
for gain is prosecuted with an earnestness that would
have borne golden fruit in any city in the Western
World. Public buildings, almost pretentious in size
and design, rise from the earth in a few days—a
residence for the Governor of Ceylon; one for the
Government agent of the province; and another for
the delegate of the Colonial Office. Amongst other
buildings are to be found a court-house, treasury,
hospital, prison, telegraph-office and post-office and
a fair example of that blessing of the East, known
as a rest-house.

Sites on the principal streets are leased for the
period of the fishery, and for ten or twelve weeks
Marichchikkaddi is one of Asia's busiest marts. One
would hardly think that these Easterners, squatting
on mats in open-front stalls, could judge the merits
of a pearl, yet they can estimate the worth of a gem
with a wonderful precision. Usually they have learned
by long experience every "point" that a pearl can
possess, knows whether it be precisely spherical,
and has a good "skin" and a lustre appealing to

connoisseurs. A metal colander or simple scale enables them to know to the fraction of a grain the weight of a pearl, and experience and the trader's instinct tell them everything further that may possibly be known of a gem.

When fishing is at its height, the scene on the banks is one of extreme animation. Each craft is a floating hive of competitive noise and activity. All around are disappearing and reappearing seal-like heads. By noon most of the divers are tired out and, if it has been a successful day, the boats are fairly loaded up. A gun is fired from the Master Attendant's ship, and this gives the signal for pulling up the anchors, hoisting the sails and beginning the run home. The men, other than the tired-out divers, occupy themselves nominally in picking over their oysters, throwing away stones, shells, and other useless things, and in preparing the loads for easy transport from the boats to the shore. But, as a matter of fact, it is well known that this opportunity is seized to "pick" the oysters in another sense. Almost invariably the finest pearls occur just inside the edge of the shells, and may fall out at any moment. No doubt many of these round and best coloured pearls are picked out during the run home and concealed about the persons of the boats' crew. This is one reason why the Government does not get its fair share of the pearls.

The homeward race of a hundred or so ruddy-sailed craft before a strong wind and over a tropical sea is a very pretty sight. They are orientally fantastic in colour and shape, and each deck is crowded with men and boys, with shining brown skins and brightly coloured cloths wrapped round them. Each crew strives to get in first, because—"first come is first served," and they who first dispose of their loads are the first to be free to rest. The load is counted and divided into three piles. An official selects two piles for the Government, whilst the other is divided among the divers. On their way to their houses these divers are besieged by a surging crowd of natives eager to buy from them their oysters by the dozen or by the half-dozen, or even singly. They may be observed stopping at boutiques and paying their score with oysters, extremely acceptable to the shopkeeper itching to try his luck. In a small way oysters pass current here as the equivalent of coin.

In the meantime those oysters belonging to the Government have all been counted and at sunset a great auction begins, the Government agent being seated at a platform looking after the proceedings. He announces how many of the oysters are for sale and puts these up by the thousand. Any number of thousand, from one to fifty or even more, are taken by individual purchasers or by syndicates.

A CEYLON PEARL FISHERY

The competition to fix the price of the first lot takes about a minute. The prices in a single night vary considerably and inexplicably; a high price, say 35 rupees per 1000, may be given at the beginning of the evening, and later not more than 22 rupees can be extracted. There is a keen and zealous competition, the larger buyers competing against the smaller, or all combining in a ring against the Government auctioneer. The day's catch is generally sold within the same night, but if not, the balance is disposed of privately the next morning.

Early the next day each purchaser comes to the Government agent with an order for the number of oysters knocked down to him the previous night, and sets to work to remove them to his own shed.

The washing of the pearls from the oysters is a most tedious, primitive, and somewhat disgusting process. The oysters are simply left to rot, the process being much assisted by vast numbers of a species of blow-fly, which after the first day or two infests the whole camp. The maggots of this fly eat their way through everything. After a week's rotting, the seething and disgusting mass is sorted by hand and the pearls, or such of them as are of sufficient size, are picked out. The residue is now ready to be washed. This is carried on in dug-out canoes or "ballams." The bivalves are put in and water is poured over them. As the water rises, a wriggling mass of maggots floats up from the lower recesses. The shells are rinsed, and the valves separated and rubbed to remove any detritus in which a pearl might lodge. The men scrutinize the nacreous lining for attached or shell pearls; placing any found in a special basket. After the quantity has been reduced somewhat the floating maggots are skimmed off. Some of the water is baled out through a sieve, any material that remains therein being carefully returned to the ballam lest a pearl may be contained or entangled in the dirt. More water is then added and the process of washing the shells is continued. Finally, after all the shells have been removed a fresh supply of water is poured into the ballam until it overflows. By this method the lighter filth is got rid of. The remainder of the water is decanted and the heavy *débris* in which the pearls are mingled is exposed at the bottom. More water is added and the detritus or "sarraku" kneaded and turned, over and over again. This "sarraku" is sorted and winnowed at leisure, and examined till the smallest sized pearls have been extracted. The final search is carried on by women, and it is amazing to see what a large quantity of small pearls their keen eyes and fine touch enable them to obtain.

After the pearls are picked out, it is customary to offer the apparently exhausted dirt for sale, and

A CEYLON PEARL FISHERY

ready buyers can always be found. In this manner the pearls of our dainty necklaces and engagement rings are wrested from Nature. The whole process is intensely interesting and picturesque, but it leaves much to think about afterwards and much to hope for. The thing has been going on in the same way for centuries, and it would continue for centuries to come if the busy Western mind, so full of new ideas and plans, were not turning its attention to improving the old system. What is wanted is to make this harvest of the sea more regular in its occurrence, to economise the vast expenditure of human energy now wasted in bringing up the oyster from the depths of the sea. The pearls might be extracted from the oysters with greater rapidity and certainty. More hygienic methods should be employed and there should be more assurance that the Government of Ceylon obtains its fair share of the revenue.

CHAPTER VII

PEARL FISHERIES OF OTHER LANDS

The Persian Gulf. The fisheries of the Persian Gulf, like those of Ceylon, have been exploited for very many years. They were known in the time of

Alexander and referred to by Pliny. By some, these fisheries are supposed to be the most valuable in the world and Seurat states that their revenue is estimated to be 10 million francs annually.

The pearl oysters that are fished are the species *Margaritifera vulgaris* (the Ceylon Pearl Oyster) and a variety of *Margaritifera margaritifera* (*persica*). The latter is shipped as Bombay shell. The pearl oysters occur all along the coast of Arabia, but the banks (like the paars of Ceylon) where the fishing takes place, are situated near the island of Bahrein. The centre of the industry is perhaps Lingah, and the pearls are sometimes known as Lingah pearls, though the term "Bombay pearls" is quite as frequently given to them. The former appellation refers to the place of origin and the latter to the seaport from which they are shipped. Several thousand boats are employed in the fishery but they are usually smaller than the Ceylon boats and the crew averages less than a dozen. The method of fishing is exactly the same as that practised by the Arab divers at Ceylon.

The diver uses the same small basket of cocoanut fibre, to which one line is attached, for the oysters.

There is also a line with a heavy stone at the end of it. The diver sinks rapidly by entering the water in a vertical position with one foot in the loop of the rope at its point of attachment to the stone.

VII] PEARL FISHERIES OF OTHER LANDS 85

At the bottom, he gathers as many pearl oysters as possible, places them in the basket, and when ready gives the signal to be hauled up. The basket alone may be hauled up with the diver hanging on to it or there may be a rope round the man's waist.

The fishery lasts from June to September when the sea is at its best for this purpose. The pearls from the Gulf are not so white or as fine as those from Ceylon. Frequently they possess a yellow colour. Curiously enough, however, in Bombay these pearls seem to be preferred to the white.

The Red Sea Pearl Fishery is another ancient industry. It was flourishing in fact in the time of the Ptolemies. Two species of pearl oyster are fished, a variety of *M. margaritifera*, which is the Egyptian shell of the trade, and also the Ceylon species which is smaller in size and fished for pearls alone. Jiddah was at one time the chief seat of the industry, but now it has lost that position. The pearls are finally exported from Alexandria and are known as "Egyptians."

The Australian Pearl Fisheries are of very considerable importance and are probably the largest and best equipped at the present time in the world.

The fishing, for either mother-of-pearl shell or for pearls, is carried on at various places on the north-west, the north and north-eastern coasts.

For a quarter of a century pearl oyster fishing

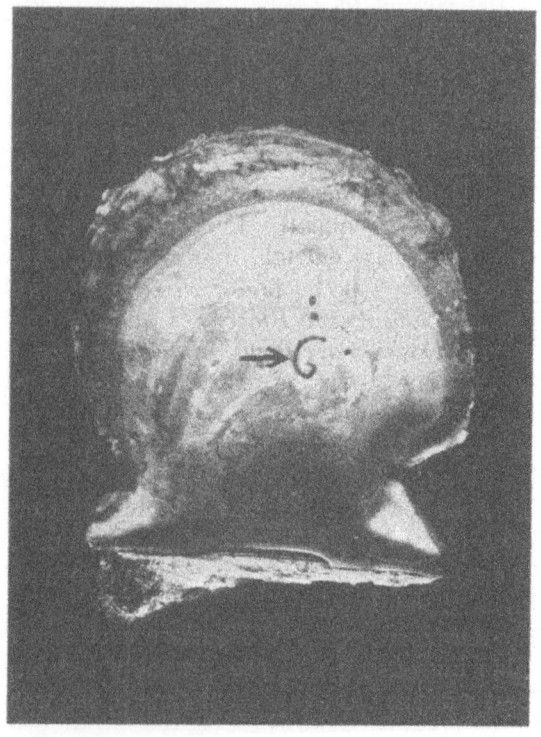

Fig. 8. Nemertine worm and sand grains entombed in a deposit of nacre.

PEARL FISHERIES OF OTHER LANDS

has ranked as one of the primary industries of Western Australia, being practised first at Shark's Bay. It has been pushed further north, and it is stated that the naked diving and the beach-combing of the aborigines financed many a settler in the early years of his pastoral enterprise.

The Queensland fisheries are carried on along the tropical coasts of that state and in particular in the Torres Strait, where some attempts at pearl oyster cultivation have been made. The Gulf of Carpentaria is another home of the industry. The custom of naked diving appears to be gradually dying or to have already disappeared in Australia, and the diving dress is usually employed. At the same time the divers are largely Malays, Japanese, and Manila-men, although the industry is financed by the Whites. As it has been always believed that it was quite impossible to have any but coloured labour, the industry has enjoyed comparative immunity from the rigorous "white Australia" laws of the Commonwealth. It was decided, however, in 1911 that this industry should conform to the laws of the country and two years' notice was given, so that this year, 1913, should see the employment of Europeans as divers. From Western Australia alone, 29,281 cwts. of pearl shell were exported in 1910, the value being £246,068. The value of the pearls discovered brought this sum up to £348,911.

The working expenses of a boat are about £500 per annum and pearl shell must be worth at least £140 a ton to bring in any reasonable profit.

The pearl fishing of *Japan* is another extensive one, and if we include China and Siberia we find that as many as 20,000 men were actually employed in fishing in 1906. Japan is the first country in the world to set up a pearl-farming industry based upon scientific knowledge. The shell fished in Japanese waters is *Margaritifera martensii* and it is particularly abundant in the bay of Agu in the province of Shima.

The district where the cultivation is carried on is also in this bay, and it is reported that as much as 22 sq. nautical miles were leased in 1911.

American fisheries. Pearl fishing is carried on on both sides of the continent of America. The Panama coast is perhaps most famous, and it is said that these pearl banks were discovered in 1530 by Vasco Nunez of Bilbao. When Spain was an all-important power in this part of the world, the country was known as Colombia and large quantities of pearls were sent to Europe.

The ancient fisheries of the Aztec kings were carried on between Acapulco and the Gulf of Tehuantepec. There are at the present time quite extensive Mexican fisheries, the shell equalling in value the pearls. The fishing seems to be systematic and new

VII] PEARL FISHERIES OF OTHER LANDS 89

beds are constantly being discovered. Both the methods of naked diving and diving with dress are employed. Another very extensive fishery is to be found on the coast of Venezuela. This also dates back before the time of Columbus. The most famous banks are on the sides of the island of Margarita which has thus received its name. About 400 sailing boats of from three to fifteen tons are employed (according to Cattelle), and in 1906 practically 2000 men were engaged in the fishing. The Government of Venezuela has given rights and concessions to individuals in return for which it is to receive 15 per cent. of the net profits. The species of pearl oyster fished is *Margaritifera squamulosa*, and it resembles the Ceylon pearl oyster. The shell is very thin and the mother-of-pearl obtained in these fisheries is worth only a fraction of the value of the pearls.

We have not exhausted by any means the pearl fisheries of the world, amongst which may be mentioned, also, the Pacific island fisheries, the fisheries of California, the African coasts and the Mediterranean. We shall conclude by a reference to the fresh-water pearl fisheries which are of importance, in the rivers of Europe and America.

The fresh-water pearl mussel of Europe is *Margaritana margaritifera*. It has been fished in Great Britain from the time of the Romans, and in the middle of the 18th century large quantities of pearls were obtained from these shellfish in the river Tay.

The Scotch pearl fishery was stimulated again in 1860. The Irish fishery seems to have died out—the shellfish being much less common now than formerly. In 1906 the value of the pearls collected from the shellfish in the British Islands was put down as £3000. No elaborate system of boats and dredges is used and probably all the shellfish are obtained by wading and feeling for the individual specimens.

On the continent the fishing has been carried on for a long time in the rivers of Finland. The pearls find their way chiefly into Russia. The fresh-water pearl mussel is also very common in the rivers of Germany and Bohemia. Quite recently the origin of pearls in these shellfish has been investigated in great detail by the Germans: reference will be made to their discoveries in a later chapter. The fishing has been very regular and carefully looked after in these countries. The rivers are inspected in springtime and the fishing takes place in summer. At the same time the recent pollution of water by factories has caused the usual decrease in the number of mussels. Pearls have also been obtained from the French rivers and in fact from many others in Europe which lack of space prevents us from mentioning.

The total value of pearls obtained on the continent of Europe in 1906 was about £20,000.

VII] PEARL FISHERIES OF OTHER LANDS 91

In America the fishery for fresh-water pearls is of greater importance, the value of pearls and shell together for 1906 being about £200,000. The fishery dates back to the time of the ancient inhabitants and remains of the shells are found in Indian mounds. The modern fishery was stimulated into growth in 1857 by the finding of a very fine pearl weighing 93 grains at Notch Brook near Paterson (New Jersey). It was sold to the Empress Eugénie for 12,500 francs.

The news of this discovery had an effect, similar on a small scale, to the discovery of gold. The American pearl mussels are found in rivers over a very wide area of the States and Canada. For further information reference should be made to the literature of American writers who have treated this subject in considerable detail.

CHAPTER VIII

THE ORIGIN OF PEARLS

WHATEVER be the cause of pearl formation in the different molluscs in which pearls are found, they are always pathological products. It may seem strange that the beautiful gems now so much in demand for ornamentation are but calcareous deposits, due to some irregularity, some perverted growth, or some abnormal process: but such is the case. There is, however, no lack of romance in connection with pearls, both in the fancies of the East and traditions of the West, and even in the sober science of to-day.

The Hindoos, whose attention must of course have been drawn very early to these glorious objects, considered (as in fact they do now) that the pearls were consolidated drops of dew. They believed that the pearl oyster came to the surface and opened wide its two valves. Drops of dew fell inside and solidified in consequence of the heat of the sun's rays. Pliny must have known of the fiction of the Hindoos when he related the tradition that the pearl oyster rested during the night with its valves open and received the drops of dew. The very quality of the pearls, their whiteness and their brilliance, were all supposed to be related to the condition of the dew-drops that had been so transformed.

Many a beautiful story has been woven around these beliefs. The pearls are Nereids' tears—again, they are dew-drops falling into the sea from overhanging leafy boughs. Other less fanciful ideas and theories have been promulgated from time to time, and every new account of the origin of pearls is introduced or prefaced by the enumeration of all previous theories.

THE ORIGIN OF PEARLS

Since this little book is written for those who are not likely to consult a long series of original works, the author may be pardoned for giving yet another similar compilation.

Author	Theory of cause
Aelian	Flashes of Lightning.
Rondeletius (1554)	Parasites and concretions.
Redi (1671)	Grains of sand.
Various 16th cent. writers	Eggs of the mollusc.
Reáumur (1717)	Solidification of shell forming fluid.
Home (1826)	A modified view of the egg theory—i.e. abortive ova.
Filippi (1852)	A worm parasite, Distomum.
Küchenmeister, Von Hessling, and Meckel (1856)	Parasites, sand, eggs.
Moebius (1857)	Entozoa.
Kelaart (1857–59)	Sand, diatoms, eggs and parasites.
Garner (1863)	Distomum parasite.
Comba (1898)	Parasites.
Dubois (1901, 1903)	Parasites.
Jameson (1902)	Parasites.
Herdman and Hornell (1902–06)	Parasites, etc.
Seurat (1902)	Parasites.
Giard (1903)	Parasites.
Boutan (1904)	Parasites.
Rubbel (1911)	Shell substance
Jameson (1912)	Shell repair substance and perhaps parasites.

It will perhaps be startling even to scientists, who have not read up the older biological literature, to see how often the theories of sand grain and the parasite causation have been brought forward in the past. The grain of sand theory dates back much further, too, than Redi, 1671.

Whilst this view dominated our older text books, and particularly school books and popular works, until quite recently, the idea that a parasite provided the stimulus for pearl formation has been repeatedly put forward since 1554. The older workers did little more than make guesses, but Filippi as far back as 1852 actually discovered a parasite as cause in the fresh-water mussel. It will be seen, therefore, that all the workers of the last ten years who have pushed the research upon pearl formation as far as fresh advances in biological technique would allow, have had at least numerous helpful suggestions left as a legacy by their predecessors.

We shall only refer in detail to one or two of the causes tabulated. First let us look at the old grain of sand theory, which modern writers have tended to throw over completely, notwithstanding the fact that Herdman and other workers have recorded sand grains as occasional nuclei in pearls.

This theory assumed that grains of sand found their way between the shell and the mantle of the mollusc and irritated the latter so that it responded by coating the invading object with layers of pearly substance. This certainly does take place sometimes,

VIII] THE ORIGIN OF PEARLS 95

and a similar process follows the introduction by man of small objects between shell and nacre-secreting tissues. As we have seen, however, in another place (p. 53) the pearly objects so formed are not really pearls, but blisters. The Chinese, for example, insert rows of images of Buddha between the mantle and shell and allow them to remain until coated with nacre. There are, however, a few cases on record where free pearls have been found with sand grains as nuclei, consequently we must assume as an occasional true cause of pearl formation—the minute sand grain. Probably this cause is of rare occurrence and of little importance.

Now let us look at two very important theories, and first at that which regards the entrance of a parasite into the shellfish as the cause of pearl formation. We have seen that the structure of pearls is intimately related to the structure of the mollusc shell. The latter is secreted by a layer of cells which bounds the surface of the mantle. Now it must be borne in mind during the following study that pearls are formed inside a little sac of cells. This "pearl sac" was recognised by von Hessling in 1858. It is composed of a single layer of cells, which are very similar to those bounding the surface of the mantle against the shell (see fig. 9). In fact, just as the mantle cells secrete shell substance of one kind or another, so do the pearl sac cells secrete the layers which build up the pearl. It will be quite clear then that if we can say definitely in *every* case how and why the pearl sac has been formed, we shall have solved the problem of the origin of pearls.

The Italian scientist, Filippi, is usually credited as being the originator of the view that the nucleus of pearls is an entombed parasite.

This investigator, whilst examining fresh-water mussels (*Anodonta*—quite common in our ponds and streams) in the royal parks near Turin (1852) was surprised at the large number of pearls found therein. At the same time the abundance of pearls was attended by the presence of large numbers of a parasitic flat worm (*Distoma duplicatum*). Filippi next examined the pearls, which were of various ages and sizes, and found always an organic nucleus. It was apparently impossible to recognise definitely in the entombed organic matter the form of the worm, but arguing from the abundance of both pearls and worm parasites (and the organic nuclei) Filippi did not hesitate to suggest that the worm parasites had caused pearl formation. Later, Filippi recognised certain parasitic remains in other pearl nuclei—remains of other species of worms.

In 1856, Küchenmeister found in both *Anodonta* and *Unio* (both common shellfish of fresh-water ponds and streams in the British Isles) the remains of an arthropod in the nuclei of pearls.

THE ORIGIN OF PEARLS

In 1857, the German zoologist Moebius took up work on pearls and penetrated much more deeply than his predecessors into the mysteries of pearl formation.

He did not content himself with the fresh-water pearls from his own country but examined those from Indian and American molluscs. Once more, the general absence of the grain of sand as nucleus was noted, and yet again the presence of a worm parasite was put on record. We find about the same period that Dr Kelaart, a medical officer in Ceylon, who was one of the first workers on the Ceylon pearl oyster, repeated the statement that minute parasitic worms were the cause of pearls. This time—that they were the cause of the true oriental pearls. Numerous investigators followed, but there is unfortunately no room here for a history of their searchings and theories, and we must skip over the intervening years until we come to the much more definite and recent work of Dubois, Diguet, Jameson, Boutan and Herdman. So far, it will be seen that workers had set themselves to discover the contents of the nuclei of pearls, and it may be admitted that it is by no means an easy task to make out in sections what the organic *débris* of such nuclei might once have been.

Now, though we are at present dealing with the history of the parasite theory, it will be of interest to note here that objections of remarkable force were made to it by von Hessling, who wrote in 1858.

The importance of his statements, however, is only just being recognised. Von Hessling believed that in rare cases foreign bodies of various kinds getting into the mollusc might provoke pearl formation, but he regarded as the chief cause some sequence of events intimately connected with the phenomena of shell growth. He makes, in this connection, the very noteworthy statement that *the nucleus of pearls is very often formed of granules of the substance which forms the external layer of the shell—the horny periostracum.*

Reference will be made to this observation later, and it will then be seen how closely this agrees with some recent views. The modern period of activity in pearl research may for our purpose be taken as dating from about 1897. In 1899, Léon Diguet described the pearl sac, already seen by von Hessling, and this was supposed to secrete nacre round the remains of a parasite which then formed the nucleus of the pearl.

In 1901, Raphael Dubois published a paper on pearl formation in Mytilus (the common edible mussel), his specimens being taken from the French coast, and during the same year Jameson followed Dubois in working at the same problem at the same place—Billiers in Brittany.

THE ORIGIN OF PEARLS

His paper followed in 1902, and confirmed the discovery of Dubois. Both these workers definitely established the fact that in the edible mussel from Billiers, pearl formation was caused by larval worms which were parasites in the shellfish. Moreover, the larvae were recognised as the young of the flatworms, known as trematodes (related to the familiar species which occurs in the liver of sheep and causes liver rot). Dubois simply followed up the methods of previous investigators. Jameson gave a much more detailed account of pearl formation, and pointed out the great importance of discovering the method by which the larval parasite caused the formation of the pearl sac. In addition to this he attempted to trace the young parasite to the adult worm. Like many other parasites, this worm requires several hosts in order that it may live its normal life. Just as the trematode parasite (the liver fluke) of the sheep requires two hosts, a pond snail and the sheep, and as the well-known malaria-causing parasite requires both mosquitos and man, so does the worm parasite in the mussel require at least one other host. Jameson attempted to piece together by experiments this life-history, and came to the conclusion that the adult form lived in a bird, the common Scoter, which feeds on the mussels and destroys large numbers of them at Billiers. His investigations gave the following interesting life-cycle for the parasite.

The adult worm (which he considered to be *Leucithodendrium somateriae*) lives in the Scoter (*Oedemia nigra*) and there produces its eggs. The eggs pass out of the bird to the external world, where probably most of them die. The fortunate ones, however, either themselves become, or give rise to a tiny free-swimming larva, which reaches the bivalve, *Tapes*, or else the common edible cockle (*Cardium edule*). In one or other of these molluscs the eggs or larvae commence the first part of their life and develop into spherical cysts. From the walls of these cysts other larvae are formed. These leave the cockle (or *Tapes*) and if fortunate eventually reach a mussel.

They bore into the mussel and come to rest in its tissues, and round them an epithelial sac of cells may appear. Now, if the mussel should be eaten by a bird—the Scoter—the larvae reaching their final host develop into the adult worm, and that is the end of the story.

If, however, the epithelial sac is formed and the mussel does not fall a prey to some predaceous bird, the trematode larva dies and round its remains the epithelial sac secretes a mausoleum of pearl. A more startling story could hardly be imagined, but most of the life-cycles of animal parasites are extraordinarily exciting. We may tabulate the stages as follows:

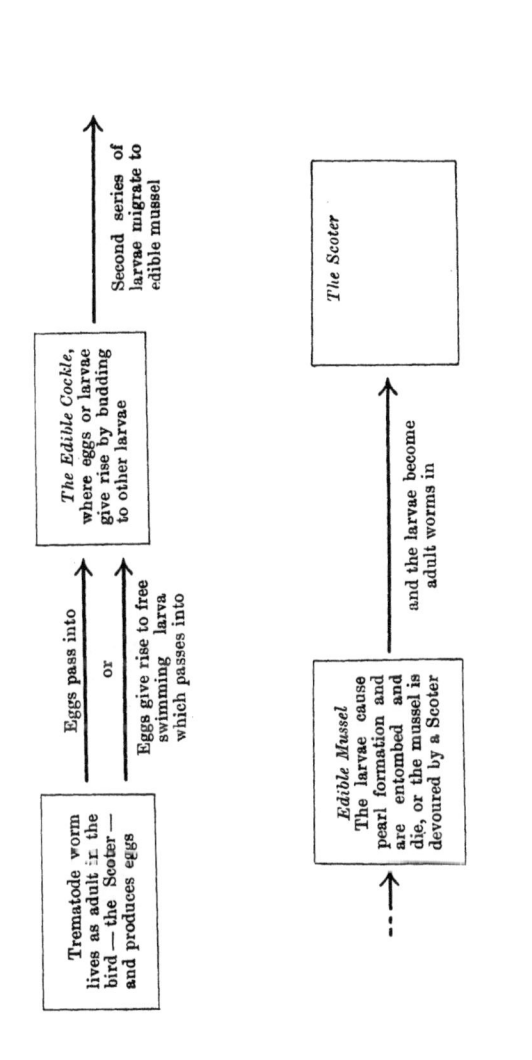

With regard to the pearl sac which appears around the worm larva in the mussel, Jameson accounted wrongly for its origin in 1902. He states that "This epithelium appears to arise quite independently of the outer epidermis, and is no doubt due to a specific stimulation on the part of the parasite, as other parasites, e.g. sporocysts, cestode larvae, etc. are not surrounded by such a sac." He continues by saying: "At first a few cells appear which proliferate and arrange themselves along the wall of the cavity.... From the first these cells are basally continuous with fibres of connective tissue. Their transformation into the pearl sac is a gradual one, and every step can be traced in sections of the parasites *in situ*." It was the Frenchman, L. Boutan, in 1904, who continued the investigation of pearl formation in the edible mussel (Mytilus) and completed the story by describing how the pearl sac was really formed by the parasite.

Boutan states that the parasitic larva in the common mussel reaches the space between the shell and the mantle of the mollusc. It then comes to lie in a little pit or depression in the epithelium of the mantle (diagram, fig. 9 *a*).

Gradually the parasite sinks deeper and deeper (fig. 9 *b* and *c*), the epithelium of the depression becomes a little flask and finally a sac just communicating with the surface by a small opening. In the end, the sac becomes cut off altogether, and we

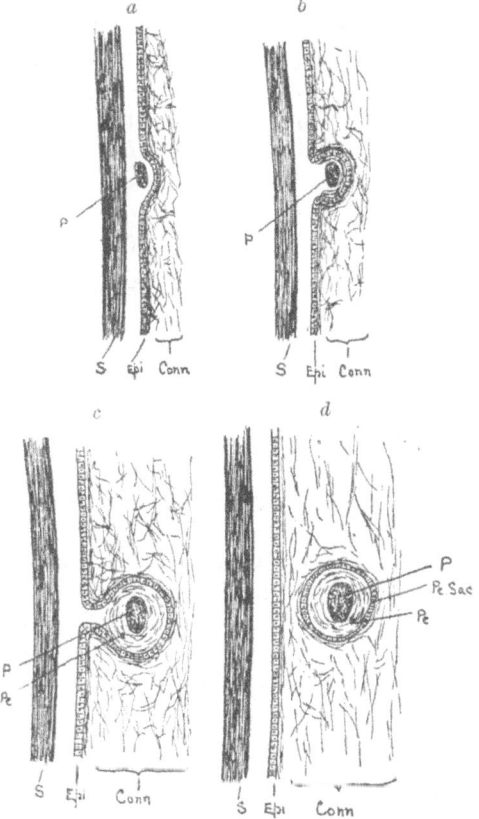

Fig. 9. Diagrams (modified from Boutan) showing formation of pearl and origin of Pearl Sac in Mytilus. S., shell. Epi., shell-secreting epithelium of mantle. Conn., connective tissue of mantle. P., parasite. Pe., pearl. Pe. Sac, pearl sac.

have a parasite lying in a little sac (the pearl sac) of epithelium which is decidedly the same as the shell-secreting epithelium of the mantle, because it was once continuous with it (fig. 9 *d*). This pearl sac will have already commenced secreting layers of shell-like substance round the parasite which has been the cause of the sequence of events leading to its formation.

Thus pearls are formed in the common edible mussel. This method, however, is not to be supposed for a moment as applicable to other molluscs without experimental evidence. As a matter of fact, it is not the *only* method of pearl formation even in Mytilus.

Now let us turn to the Ceylon pearl oyster which produces such valuable pearls and see what processes are involved there.

The problem of the origin of the valuable oriental pearls is obviously of greater economical importance than that of the Mytilus pearls. The discussion of the theories brought forward will be given in two separate sections, for if the description of pearl formation in the fresh-water mussel is interpolated in the middle of our account of the Ceylon pearl oyster we shall be following the discoveries in their proper sequence.

In 1902 the British Government sent to Ceylon a scientific commission (consisting of Prof. Herdman and Mr J. Hornell) in order to report upon the

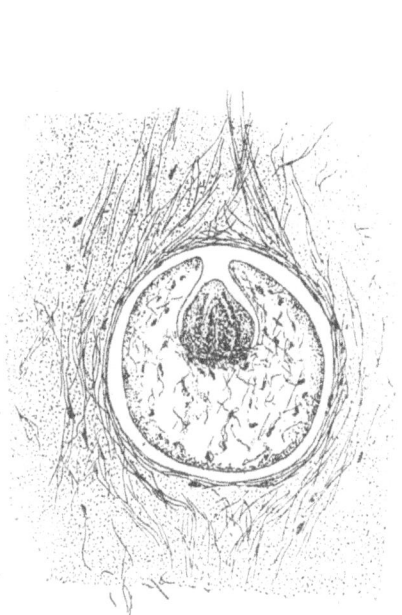

Fig. 10. Section of larval worm (Cestode) parasitic in Ceylon Pearl Oyster. (After Herdman.)

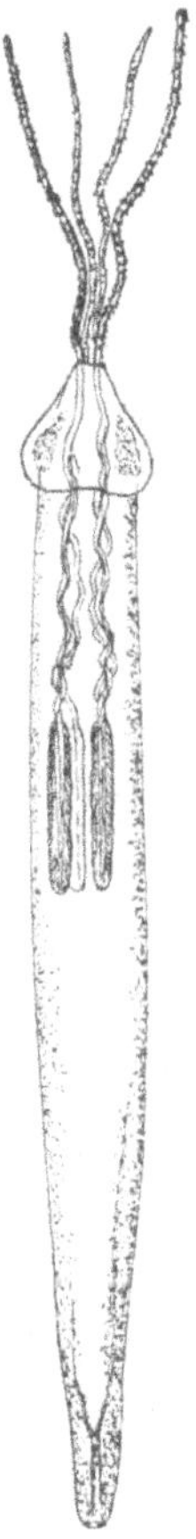

Fig. 11. *Tetrarhynchus unionifactor*. Practically young adult stage of one of the parasitic worms of the Pearl Oyster.

condition of the famous pearl banks of the Gulf of Manaar, which, notwithstanding numerous bad years, have yielded much profit to the Government of Ceylon. The results of their investigations[1] are extremely valuable, and it must be remembered that the primary object of the expedition was to report upon the conditions of the pearl banks, the distribution of the beds of pearl oysters and the influences affecting their life and prosperity.

However, in addition to this work, Herdman and Hornell also examined many pearl oysters and such pearls as were obtainable for the purpose of determining the cause of their formation.

Obviously they were well acquainted with the numerous theories of their predecessors when they first went out to Ceylon, and consequently when in March 1902 they noticed numerous white larvae (fig. 10) parasitic in the "liver" of the pearl oyster, they devoted some attention to their occurrence and nature, and came to the conclusion that these parasites were the young of cestode worms (the most familiar cestode worms are the common tapeworms of man, the dog, etc.), and subsequent work showed that some at least of them were the larvae of a worm that occurs frequently in Elasmobranch fishes and is called *Tetrarhynchus* (fig. 11).

[1] Given in a Report to the Ceylon Government, published by the Royal Society, in 1903–6.

THE ORIGIN OF PEARLS

Now, sections of the Ceylon pearls *in situ* differ considerably from those of Mytilus pearls, and in two ways. The pearl sac is not nearly so distinct, and the nucleus of the pearl is much smaller and looks different. In several cases Herdman could not find any nucleus at all.

We have already seen that there are two kinds of pearls in the Ceylon pearl oyster, "cyst" (or parenchyma) pearls and "muscle" pearls, the latter occurring in large numbers near the insertions of the muscles. The muscle pearls were *never* believed by Herdman to have been caused by parasites. Herdman and Hornell stated that it was probable that these pearls were formed by the deposition of calcareous matter round some minute calcareous particles or calcospherules. The origin of the calcospherules themselves was guessed at by Southwell, who stated that they were almost certainly depositions from the blood.

From what we have already seen, we should expect an epithelial sac to be present to deposit this calcareous matter. From whence does this come if not brought in by a parasite? Herdman and Hornell, noticing that the muscle pearls were formed very close to the surface of the mantle, found no difficulty in presuming that the epithelial cells on the surface of the mantle might migrate inwards to the source of irritation. So much then for the muscle pearls to which we shall return later. Suffice it to say that Herdman and Hornell believed them to be formed owing to *internal causes* and not to any parasite or other body of external origin.

With regard to the other really fine oriental pearls, Herdman and Hornell came to the conclusion that the chief agent concerned in their formation was the tapeworm larva, which occurred so frequently in the pearl oysters. The oriental pearls are of course like all others formed by the activity of a pearl sac, and Herdman recognised the great resemblance of this to the outer shell-secreting epithelium. The origin of the pearl sac has, however, never been observed in the Ceylon pearl oyster, but Herdman thought that the parasite boring into the mollusc might carry in certain epithelial cells from the surface.

What is the life-story of this parasite which occurs in the pearl oyster? It must be confessed that so much doubt exists concerning the relation of the encysted larvae to both the pearls and the adult tapeworms that the technicalities are beyond the scope of this little book. The conclusions of Herdman, Hornell and Shipley (who described the parasites found in the Ceylon pearl oysters and fishes) can be expressed in the following form:

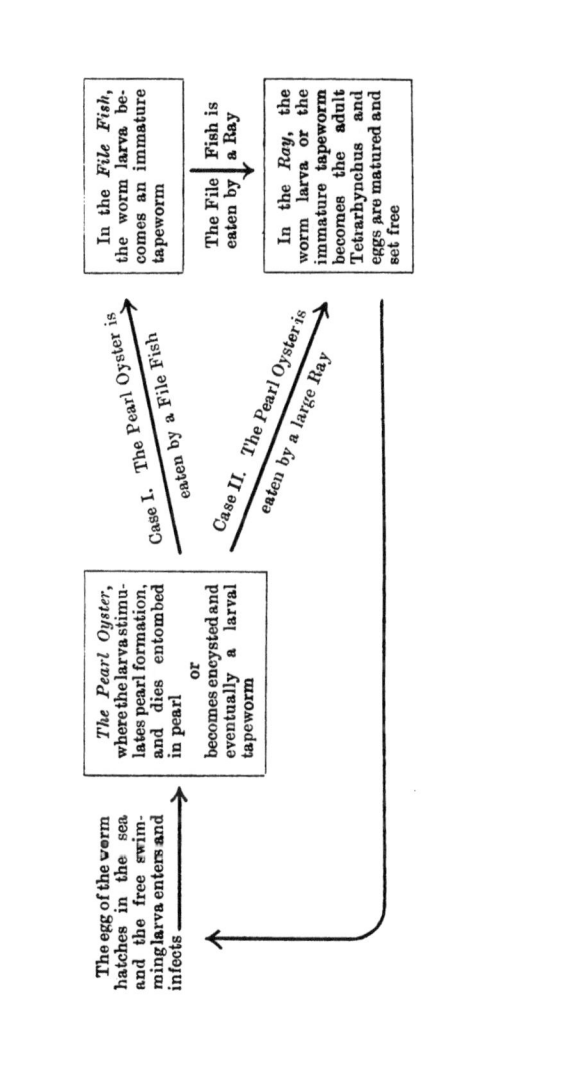

It will be seen that there are two possibilities. If one be taken there are two hosts only for the parasite. In the other case there are three hosts. In this latter, the worm larva has first to enter a pearl oyster (in which it passes its first days). The pearl oyster must then be eaten by a file fish or other Teleostean in order that the second stage of life of the parasite may be safely negotiated, and finally a third host, the ray, must be reached, by the annihilation of the file fish:—truly an interesting picture of the dependence of animals one on another. If these theories are correct, it is therefore necessary for pearl production to be assured to have pearl oysters, worm parasites, file fishes and the rays, present together in the same locality.

As a matter of fact it is very probable that the file fish is only an accidental host and that the direct passage of the larva from the pearl oyster to the ray is all that is required[1].

Now, are we to consider that the majority of fine oriental pearls have been produced by the influence of worm parasites or not? Jameson doubts the cestode worm theory, although in 1902 he stated that he himself found worm parasites in decalcified and sectioned pearls from the Ceylon oyster. Let us

[1] Jameson believes that there are three distinct species of parasites encysted in the pearl oyster, but in each case the adult worm is to be found in some fish that feeds upon the oysters.

THE ORIGIN OF PEARLS

leave the Ceylon pearl oyster for a while and turn to the most recent scientific work on the fresh-water mussel in Germany.

Rubbel, working at Marburg, has once more shown how often the older scientists came near to the solution of problems not finally solved till to-day.

According to the investigations of this worker the fresh-water mussels (*Margaritana margaritifera*) examined by him contained few parasites that *could* give rise to pearls. Parasites were, moreover, *not* found to be the cause of pearl formation in this shellfish. The nuclei of the pearls consisted of particles of a yellow-brown substance, that was strongly refractive. This substance could be found scattered in the connective tissue of the mantle and also in the shell-secreting epithelium of the mantle. Furthermore it seemed to be related to the substance of the outer layer of the shell, i.e. the periostracum. In short, we are reminded of the views of von Hessling, put forward in 1859.

In order to understand fully the mechanism of pearl formation, Rubbel first made experiments to determine what the outer shell-secreting epithelium of the mantle could do in the way of repairing broken shells or regenerating pieces of shell that were removed. He discovered that the nacre-secreting epithelium, normally secreting the mother-of-pearl layer, had the power or potentiality of producing all the other layers of the shell.

The pearls, however, as we have already seen, consist of some or even *all* of the layers of substance which form the shell. They sometimes consist only of periostracum and are yellow or brown and horny looking in appearance. Most pearls, however, consist of concentric layers of periostracum, prismatic substance and nacre, with another layer present in places—the hypostracum. These layers may alternate irregularly (figs. 6 and 7).

The yellow-brown granules, referred to above as being akin to the substance of the periostracum, are to be found in the outer epithelium (fig. 12 *a*). Rubbel believes that some of them are dissolved; others, however, remain and stimulate the epithelium in some way so that they become surrounded with cells directly continuous with the outer epithelium (fig. 12 *a*). There are many points of resemblance here with the effect of parasites as described by Boutan, but it will be noticed that the granules (internally produced) are the cause, and that they do not lie outside the epithelium. The cells surrounding the granules secrete new substance round this nucleus, and we may already speak of the little body so formed as a young pearl. Gradually the epithelial pearl sac with its contained pearl breaks away from the outer epithelium (fig. 12 *b*), until finally it is completely

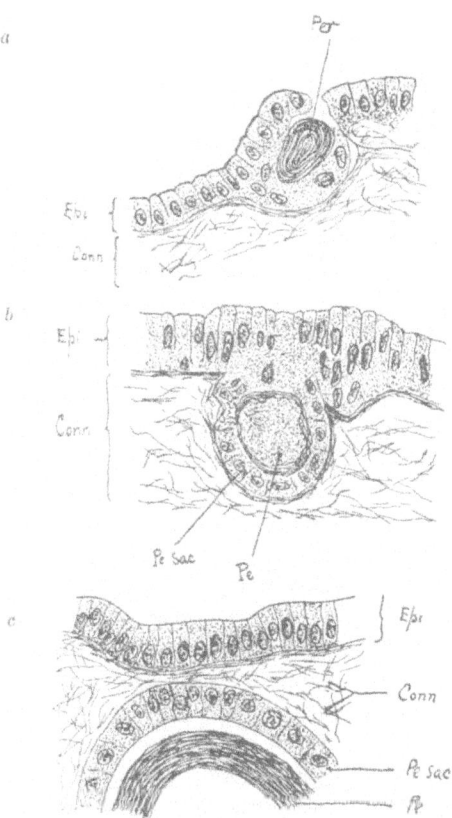

Fig. 12. Figures showing the development of the Pearl Sac according to Rubbel (modified from Rubbel). Per., periostracum-like substance forming the nucleus of the pearl. Epi., epithelium of mantle. Pe. Sac, pearl sac. Pe., pearl. Conn., connective tissue of mantle.

separated and lies free in the tissues (fig. 12 c). So the pearl formation in this fresh-water mussel commences actually in the outer epithelium and is caused by a product of the mollusc itself. Thus we arrive at a very important point, which is that there is no obvious difference between "muscle pearls" and other true pearls, for both are formed round nuclei which are products of the shellfish. Any difference in structure may be simply due to their difference in position.

The pearl sac appears to possess the power of secreting all the various kinds of shell substance, one after the other. Parts of it may even be secreting one kind of substance whilst other parts secrete another. What determines the character of the substance to be secreted is quite unknown.

Thus we have dealt with a second method and cause of pearl formation, which has now been demonstrated by Rubbel to be of frequent occurrence.

In conclusion let us return once more to the oriental pearls from the Ceylon pearl oyster. We have seen that Herdman recognised various causes, and that the muscle pearls were supposed by him to be formed round a substance which was a product of the shellfish itself. The chief cause, however, in the case of cyst pearls was supposed to be a cestode parasite which was found infecting the oysters in abundance.

THE ORIGIN OF PEARLS

Jameson examined a very large number of oriental pearls and published an account of his results last year (1912) which, though written without any knowledge of Rubbel's paper, agrees in a remarkable way with this author's conclusions for the fresh-water mussel. This renders it extremely probable that the method described above, and discovered by Rubbel, is of very general application, and is the chief cause of pearl formation in many other molluscs.

Jameson does not call the layers seen in the pearl by the names of similar layers in the shell. He recognises the resemblance to repair substances of regenerated shell, and so uses the terms "granular repair nacre," "columnar repair substance," and "amorphous repair substance," for the nacre, prismatic substance and periostracum respectively.

The formation of the pearl sac has, however, not even yet been traced in the Ceylon pearl oyster. From the evidence of structure Jameson concludes that the tapeworm cause is of little importance in the case of Ceylon pearls, and that whilst sand grains and other foreign particles or organic bodies may form the nucleus of these pearls, a foreign nucleus is probably exceptional. The factors giving rise to the pearl sacs have therefore yet to be discovered.

We would suggest as probable that eventually a sequence of events similar to that observed by

Rubbel in the freshwater mussel, will be discovered in the case of the Ceylon oyster.

Thus we have seen how pearls may be caused by foreign bodies (sand grains, parasites, or other objects) or by internal products of the pearl-producing mollusc. As Dubois, in a recent summary of the matter, puts it: "Il existe deux catégories de margaritose ou maladie perlière; l'une parasitaire et l'autre non parasitaire."

The importance of the causes referred to varies according to the species of shellfish in which the pearls are found, but whichever cause we look into, we find something of the romance of science, a knowledge of which cannot fail to increase the value of the pearl as a gem, in the eyes of any enlightened owner.

CHAPTER IX

PEARLS IN JEWELLERY AND TRADE GENERALLY. ODD FACTS

It will have become evident from the preceding chapters that the pearl is still retaining its hold in the fashionable world of the 20th century. As a matter of fact, the value of pearls has steadily increased in the last decade, notwithstanding the opening of new fishing grounds. The pearl industry (speaking only of the actual fishing itself) plays no small part in the economics of the world, and the following table, from Kunz and Stevenson's volume, showing the number of fishers and the approximate value of the products found by them in 1906, supports this statement.

	Pearl fishers	Local value of pearls £	Local value of shell £
Asia.			
Persian Gulf	35,000	800,000	22,000
Ceylon	18,500	240,000	8,000
India	1,250	20,000	19,000
Red Sea, Gulf of Arabia, and African Coast	3,000	40,000	30,000
China, Japan, Siberia	20,000	80,000	10,000
Europe.			
British Isles	200	3,000	
Continent	1,000	20,000	600
Pacific Islands.			
South Sea	4,500	25,000	100,000
Australian Coast	6,250	90,000	240,000
Malay Archipelago	5,000	60,000	160,000
America.			
U.S.A. rivers	8,500	130,000	70,000
Venezuela	1,900	55,000	2,000
Mexico	1,250	42,000	40,000
Panama	400	8,000	15,000
Various	1,000	15,000	5,000

It would be quite impossible to make any statement as to the value of pearls, for so many details enter into the question. It is primarily judged by size or weight, but this may be modified altogether by the shape, lustre, colour and perfection of the pearl generally. One or two details of interest may, however, be given. In the first place, the value of pearls is reckoned not on the weight, but on the square of the weight.

The pearl grain, which is a fraction of a carat, is the unit, and a base price is the value of this unit. What the base price may actually be, we shall not attempt to estimate. If, however, the base value of a 1-grain pearl is taken as being 20s., then the value of a 2-grain pearl would be $2 \times 2 . (20s.) = £4$. A 5-grain pearl would be worth $(5 \times 5) . 20s. = £25$. Even this method does not hold good for very large pearls of perfect lustre. The carat and grain both vary in the different countries in which they are used, although a very representative circle of pearl merchants is now upholding a metric system carat which equals 200 milligrammes.

So far as the value of mother-of-pearl shell is concerned, the following figures may be given. The values are, however, always liable to much fluctuation and these figures are not of quite recent date. Port Darwin shell (North Australia), £8—£9. 10s. 0d. per cwt.; Queensland shell, £8—£11 per cwt.; West Australian shell, £8—£9 per cwt. (1910).

Reference has been made in the chapter on

pearls to the manufacture of artificial pearls. As the price of pearls has advanced, very much care and ingenuity has been shown in mounting and altering pearls which have defects. Surfaces may be restored by scratching or roughening. Badly coloured pearls may be bleached, but this is a dangerous method and the pearls soon deteriorate after such treatment.

Skinning (or peeling) is a very well-known operation. It consists in removing the outer layers or laminae from pearls which are not quite perfect. This, however, presupposes the presence of better layers underneath The result may be a very much enhanced value or the reverse. Skinning requires great care and it is performed under the magnifying glass with steel files.

The drilling of pearls is another delicate operation, although the introduction of special apparatus has increased enormously the number that can be handled in a given time. Much skill can be shown in choosing the point where the hole shall be drilled. Small defects can be hidden in this way just as imperfections in shape can be toned down by careful mounting.

Pearl stringing is another art. Each pearl on a string should be secured by knots so that breakage of the cord will not be followed by a scattering of precious gems. The cord used is of the very best silk, and "restringing" should take place every 3 to 6 or 12 months according to the wear that takes place.

Only this Christmas (1912—13) there appeared in the press a story of the "dying" of pearls. Many similar stories have appeared from time to time, all more or less sentimental. As, however, pearls do not live, they cannot die. Moreover, pearls well treated practically do not change in appearance. As a matter of fact, the gem is at its very best when it is first abstracted from the shellfish in which it has been formed. Pearls are sometimes stained yellow by exudations from the skin, especially if the latter is unhealthy. The same staining effect is produced by sulphurous or other polluted atmospheres.

This change in colour, or lustre, may sentimentally be described as "death," but the statement certainly has no real meaning. There seems also to be a strange lack of evidence supporting the common belief that some individuals can restore the lustre of pearls by wearing them.

There are, of course, various methods of cleaning pearls and restoring them, but perhaps it will be advisable to refrain from giving them here. Good pearls require careful handling and had better be left in the care of specialists.

This little book would be incomplete if no mention were made of some of the famous pearls in the world's history, and probably first in this respect must be placed the gems worn by Cleopatra.

Almost everyone has read or heard of the wager

made by Cleopatra with Mark Antony and its sequel. The story is both antique and we regret to say absurd. Cleopatra is supposed to have dissolved one of the pearls suspended from her ears (a pearl, the value of which has been estimated at £60,000!) in wine, and to have drunk the solution. Chemically, this is quite impossible, for the organic matter would certainly not dissolve, and very little if any of the inorganic would have been affected even by vinegar in the time. She might of course have swallowed it as a pill! The other pearl, its fellow, is supposed to have been cut in halves and placed in the temple of the Pantheon.

Charles the Bold, Duke of Burgundy 1433—1477, is said to have possessed a very large and valuable pearl. The pearls of Mary Stuart are also to be reckoned amongst the most famous of the time. They were said to be the most beautiful in Europe.

Of course many of the most precious pearls have not been brought to the publicity of the auction room and may still exist in the East. Reference may be made here to the wonderful Peacock Throne built by the same oriental king who raised the Taj Mahal. Very many pearls were used, amongst which was a choice gem weighing 200 grains. The total value of the jewels was estimated at £12,000,000, but the present value is much less, probably somewhere near £2,000,000. Perhaps the most costly ornament in the world is a shawl or carpet of pearls possessed by the Gaikwar of Baroda. The value of this has been put down as over £1,000,000.

Of the very large and fine pearls often referred to by writers, two may be mentioned here, the Hope Pearl and La Pellegrina. The last-named was one of the most wonderful pearls in the world. The weight, $111\frac{1}{2}$ grains, was extraordinary for a perfectly spherical pearl of matchless lustre. It was owned by a Russian and is said to be still in Moscow, although the writer thinks that some doubt exists with regard to this.

The Hope pearl belonged to a London banker. It is supposed to be the largest pearl known in the world and weighs 1800 grains or roughly 3 ounces. It is, however, not spherical, and is more correctly a baroque. It was sold at Christie's in 1886 and is now valued at £9000.

No one who has visited and really knows Naples, can have failed to have seen something of the cameo industry. Cameos are of course not made from pearls, but as they are carved from mollusc shells and depend upon the layers that we have referred to so many times, it will not be out of place to mention the industry here.

The shells used are chiefly univalve coiled shells related to our common whelk, but they come usually from tropical seas. The chief species employed are

Cassis tuberosa Linn., *Cassis cornuta* Linn., and the "conch" shell of the West Indies—*Strombus gigas* Linn. The latter is a very large shell but is not so valuable as the others. Large quantities were in former days imported into Liverpool from the Bahamas, but the writer can get no evidence of the continuation of this traffic. The manufacture of cameos consists in carving away certain of the surface layers of the shell. The deeper layers are differently coloured. Consequently an engraving results, in which the picture stands out on a background of another colour. In the case of the conch shell, the cameo is yellow on a beautiful red. The Cassis shells give white on an orange or golden background.

It may be surprising to most readers of this little book to hear that pearl-like structures occur sometimes in plants. It may be quite incorrect to speak of them as pearls, but some of them, especially rare specimens from the cocoanut, are said to resemble in colour and appearance the real pearls from the sea.

It appears, however, most difficult to get any conclusive evidence about these calcareous concretions.

Riedel (according to Professor Korschelt) is supposed to have found one of these in a cocoanut in 1886 whilst in North Celebes. It was pear shaped and 28 mm. long. Harley, who made chemical analyses of the real pearls, examined a supposed vegetable pearl that came from Singapore. He was very sceptical, however, because it resembled rather closely a pearl from the bivalve *Tridacna* and its composition was very different from the constituents of the cocoanut. Calcium is usually deposited in plants as oxalate, although of course numerous calcareous algae are known in which calcium carbonate is deposited.

CHAPTER X

THE PEARL OYSTER—THE TRADER—THE MAN OF SCIENCE—THE CULTIVATION OF PEARL OYSTERS

We have described the pearl, the pearl fishery, and the economic and scientific aspects of pearling; what should be more natural than a combination of science with the pearl industry? Science is now such a force in the world that we should expect it to come to the aid of the pearl fisher.

Before the jeweller can mount the valuable gems from molluscs of the sea or fresh water, the fisher must supply him with the materials for his trade. This means first and foremost that large beds of flourishing pearl-producing shellfish are required. Then, secondly, we require that these molluscs shall *produce* pearls in quantity and of good quality. A prolific supply of shellfish does not necessarily mean a handsome supply of pearls, and it has been noted very often in the case of the common mussel that the molluscs from some beds produce far more pearls than those from other areas. At the same time, without the molluscs there will be no possibility of pearls. Evidently then, the first care must be for the beds of molluscs.

Granted that shellfish of the right kind are present in abundance, we might turn our attention to artificial improvement of their powers of pearl-production. As will be seen in the course of this chapter it is still the cultivation of the pearl oyster that requires the first aid from science, and certainly, so far as Ceylon is concerned, we must agree with Herdman, who has regarded the condition and the welfare of natural beds of oysters as a more important problem than the question of pearl-production.

It must be confessed that so far science has not proved a marked success in its application to pearl industries. This, however, is not the fault of science, but rather its misfortune, and the latter is no doubt largely due to our treatment of biology, the science concerned, in this country.

Economic biology is of modern, very modern growth, and it has had a severe fight in many places for its existence. Pure scientists have for some reason looked askance at it, and the non-scientific public has not seen its possibilities, except perhaps in the one branch of economic entomology. The chemist, the engineer, and the architect may gain much practical knowledge in our universities which directly affects technical work. There is, on the other hand, a considerable difference between the biology of most university courses and the biology of the North Sea Fisheries or the Pearl Banks of Ceylon. The engineering student is not required or expected to build mighty bridges immediately after a university career, but the student of biology without any practical experience is often expected to solve most difficult problems and to alter considerably the normal course of nature in a few months. Added to this, however, we have the extraordinary fact that whilst engineers are appointed to superintend the construction of government railways, architects to design government buildings, and so on, apparently anyone, without any previous training, may be appointed to take care of our fisheries. All this means that in our educational work less interest is shown in providing teaching facilities which would train the men required. It is certainly rather surprising that in the country which sent out the *Challenger* Expedition, which led the way in deep sea cable laying, and which reaps a harvest of many millions a year from sea fisheries, there is practically only one university where definite oceanographical and fishery courses are held. If they were asked for, they could easily be arranged, but under the present conditions there have been few posts for scientists trained along these lines and the demand for the training has in consequence been small.

Again, the number of students who take biology as a subject is rather small, because, as a rule, zoological courses are attended by students who hope to become teachers in various kinds of institutions, or to take museum posts. Since, owing to mistaken notions, or false modesty, zoology does not occupy the place that it should, in our school curricula, educational institutions do not require many teachers with a knowledge of this subject. The fact remains that considering the income from successful pearl fisheries, there has been really little demand for scientific interference, and there have been very few individuals capable of doing such work without requiring special training.

It is impossible to say when man first endeavoured to stimulate the pearl-producing molluscs to form pearls or to cover with nacre objects inserted inside them. It is attributed by the Chinese to a native of Hou-Tchéou-Fou who lived in the 13th century.

The rearing or protection of the pearl molluscs

is more modern. Strangely enough the commercial success attending the artificial stimulation of nacre formation has been greater than that accompanying the efforts of mollusc rearing. It seems to be very difficult to maintain the natural conditions favourable to growth and breeding, and the success which has attended the cultivation of the edible oyster in parks and other places, has been conspicuous by its absence from pearl oyster culture.

In recent years Japan has advanced very considerably in combining the work of oyster rearing with operative work for the artificial stimulation of pearl production. The work has been cared for by both the scientist and the business man, and consists in collecting and cultivating the pearl oysters of Japanese waters, on grounds suitable for them. After the shellfish have reached a suitable size, a bead of mother-of-pearl is introduced in the usual place between the mantle and shell. This, in the course of time, produces a pearl-like body—or culture pearl (Jameson). These culture pearls are not, of course, real pearls, but still, the industry is of such extent as to pay its way. It is hoped to produce real pearls artificially by a modification of the method employed as above.

Several attempts have been made in Australia to interfere with the normal course of events in the pearl oyster industry. The late Mr Saville-Kent, famed for his studies on the Great Barrier reef, urged strongly the system of cultivation of the Australian mother-of-pearl oyster. He made certain experiments in Western Australia which showed that shells could be transplanted to very considerable distances from their natural habitat.

In 1906 Saville-Kent formed a company—The Natural Pearl Shell Cultivation Company, Ltd.—for the purpose of cultivating the mother-of-pearl oysters in the Torres Straits. Several thousand large oysters were laid down in the hope that they would provide young spat which would settle on the ground and grow up. Nothing seems to have come out of this work and the lease was abandoned in 1909 at the time of Mr Kent's death.

Earlier still than this, another company was at work in the Torres Straits. This was the "Pilot Cultivation Company." Their purpose (according to Jameson, who visited the Straits in 1900) was to transplant undersized mother-of-pearl oysters by ship to an area in the Straits. The company laid down over 100,000 undersized shell and three years afterwards these were fished up. Many thousands appear to have died, but about 35,000 were taken up, and it is probable that they realised a considerable sum. It is only too probable that this was overshadowed by the expenses of transplanting and collecting the shell, and experimentation.

Several attempts have been made at one place or another to raise oysters in tanks or docks so that the young spat would be saved and could be cultivated. So far, practically no signs of any success of economic value seem ever to have been in sight. The nearest approach to success was the attempt of Mr Haynes (afterwards the work of a syndicate, the Montebello Shell Syndicate, Ltd.) at the Montebello Islands. A tidal pond of considerable extent was made and several hundred oysters were laid down as breeding stock. Young oysters *did* appear, but Jameson considers them not to have been the young of the breeding stock, but to have been immigrants from the sea, of another species. Eventually these also disappeared. So far, therefore, nothing but expense has resulted from this effort.

There is no doubt whatever, as Jameson states, that the Government of Australia should very seriously consider this pearl oyster industry. It can only be improved by a thorough and detailed investigation by qualified scientists and pearl-shellers.

Finally, a glance at the scientific inquiry into the Ceylon Pearl Fishery, the results which followed and the advice tendered to the Government may give the readers of this little work some idea of the problems which have to be faced. It was in the year 1900 that the Government of Ceylon, after a period of ten barren years, began to consider

PEARLS AND SCIENCE

whether it might not be possible to prevent the recurrence of such periods. Professor W. A. Herdman, F.R.S., was selected on the advice of the Council of the Royal Society and of Sir Ray Lankester, then director of the British Museum, for the purpose of conducting a thorough investigation of the pearl banks. He took with him as an assistant, Mr James Hornell, and left England at the end of 1901 for a visit to the Gulf of Manaar.

During the three months of Herdman's sojourn at the island, cruises were made on a steamer fitted for the investigation and all the principal pearl banks were examined. Dredging, trawling and diving were carried out, and the biologists also took part in the annual government inspection of the pearl banks.

As a result of this expedition the condition of the pearl banks was made known in great detail. Collections were made of the animals and plants living along with the pearl oysters. Some of these are merely passive neighbours, others are active enemies or competitors for food. It was necessary to determine what factors influenced the pearl oysters in their growth in order, if possible, to arrive at the causes of failure of fisheries. Herdman's reports contain a mine of information, and we must say that the work which the expedition was sent to accomplish, was carried out thoroughly and successfully.

After Herdman returned to England, Hornell

remained in Ceylon to continue the work and was later (1904) appointed Marine Biologist to the Government.

There is no need to give more than a brief summary of the chief points of Herdman's conclusions from his preliminary report (1902) to the Government of Ceylon. They are as follows:

"1. The oysters we met with seemed on the whole to be very healthy.

2. There is no evidence of much disease of any kind.

3. A considerable number of parasites, both external and internal, both Protozoan and Vermean, were met with, but that is not unusual in molluscs, and we do not regard it as affecting seriously the oyster population.

4. Many of the larger oysters were reproducing actively.

5. We found large quantities of minute "spat" in several places.

6. We also found enormous quantities of young oysters a few months old on many of the Paars. On the Periya Paar the number of these probably amounts to over a hundred thousand million.

7. A very large number of these young oysters never arrive at maturity. There are several causes for this:—

8. They may have natural enemies, some of which we have determined.

9. Some are smothered in sand.

10. Some grounds are much more suitable than others for feeding the young oysters, and so conducing to life and growth.

11. Probably the majority are killed by overcrowding.

12. They should therefore be thinned out and transplanted.

13. This can easily and speedily be done, on a large scale, by dredging from a steamer, at the proper time of year, when the young oysters are at the best age for transplanting.

14. Finally, there is no reason for any despondency in regard to the future of the pearl oyster fisheries, if they are treated scientifically.

The adult oysters are plentiful on some of the Paars and seem for the most part healthy and vigorous; while young oysters in their first year, and masses of minute spat just deposited, are very abundant in many places.

"To the biologist two dangers are, however, evident, and, paradoxical as it may seem, these are *overcrowding* and *overfishing*. But the superabundance and the risk of depletion are at the opposite ends of the life-cycle, and, therefore, both are possible at once on the same ground—and either is sufficient to cause locally and temporarily a failure of the pearl oyster fishery. What is required to obviate these two dangers ahead and ensure more constancy in the fisheries is careful supervision of the banks by someone who has had sufficient biological training to understand the life-problems of the animal, and who will, therefore, know when to carry out simple measures of farming, such as thinning and transplanting, and when to advise as to the regulations of the fisheries.

(Signed) W. A. HERDMAN."

In Herdman's further Report of 1904, certain recommendations were made to the Government as a result of his inquiry.

He suggested that the dredge should be used more frequently, to supplement diving for oysters, and that more attention should be paid to inspecting other regions of the pearl bank plateau than the known paars. Furthermore, that all young oysters appearing on certain paars, where they never reach maturity, should be transplanted to more favourable grounds. Another important point was that an attempt should be made to increase the area available for attachment and growth of young pearl oysters, by artificial "culching," that is to say, scattering dead coral, rocks or rubble on the sandy bottom adjoining the more important paars.

The following agents were recognised as causing widespread death of the pearl oysters:

1. Shifting sand, due to strong currents;
2. Voracious fishes, chiefly rays and file fishes;
3. Invertebrate animals, such as boring shellfish, boring sponges, starfishes and the smothering mollusc, *Modiola*.

At the time of this scientific investigation, it was seen that certain banks were covered with pearl oysters and there was a prospect, if all went well, of a fishery in 1903. This fishery came off and was the first for 12 years. The revenue netted by the Government was over £55,000

The following years, 1904 and 1905, were still greater successes and record fisheries took place, giving £71,150 and over £167,000 respectively.

Such was the position, when in 1906 a company was formed and the Ceylon pearl fisheries were leased for a period of 20 years at an annual rental of Rs. 310,000. In addition to this, however, the company had to undertake to spend from Rs. 50,000 to Rs. 150,000 annually "on the experimental or practical culture of the pearl oyster and on the improvement of the pearl banks within the conceded area, and the maintenance thereon of a proper breeding stock, or otherwise, in the improvement and development of the fishery." Now it must be noted here that science never advocated the formation of such a company with the fundamental mistake of having to pay such heavy charges. The company had, in fact, agreed to pay the Government a rent which was about three times the average annual revenue! This might have been reasonable if the fisheries of 1904 and 1905 had been certain to be continued; but there was no evidence that Herdman's recommendations had been carried out to such an extent as to warrant this belief, nor could it ever have been expected that the scientific recommendations implied for the future an entire absence of barren years.

The Ceylon Company of Pearl Fishers, Ltd., entered, therefore, upon a highly speculative concern, although it is quite possible that they expected science to pull them through. They were successful in the fishery of 1906, and another fishery followed in 1907. Thus the first years of the company were marked by such handsome returns that a very good dividend was paid within the first few months of the company's existence. In fact, about £50,000 was paid in dividends (during the first two years) on an issued capital of £90,000! Since 1907, however, there have been no fisheries and the company paid rent to the Government until lack of any prospects of a fishery resulted in a re-arrangement of terms and the termination of the lease in April 1912.

Statements have been made that the company followed absolutely the advice and recommendations of the scientific advisers, but it seems quite doubtful how far from a scientific point of view this was the case. The recommendations were apparently not carried out to anything like an adequate extent; and sufficient adult pearl oysters to form a breeding reserve were not left at the end of the last fishery.

So far as I am aware, the points put forward by Herdman in his report are not in the least disputed, notwithstanding the years of observation since 1902. Possibly the part played by voracious fishes in devastating the oysters was underestimated, but still Herdman recognised them as being one of the most serious enemies of the pearl oyster. He laid great emphasis also on the danger of overfishing which seems to have been sometimes carried on. There is absolutely no doubt that this commercial failure is not attributable to biological science, and Herdman himself did not advise the formation of the company.

It seems to the writer that the criticism that the scientific advisers should have devoted more time to the study of the mechanism of pearl formation is quite unfair. What was required from them was a study of the pearl oyster and the condition of the pearl banks. Without the oysters, it is obviously quite impossible to get pearls at all.

Commercial pearl fishing still remains a great speculation. It will require further very careful study and much more experiment undertaken in accord with scientific advice before nature is reined in or guided along paths suitable to man's industries.

BIBLIOGRAPHY

Only a few references to literature are given here. More complete lists can be obtained from the memoirs referred to below.

General

KORSCHELT. Perlen. Fortschritte der Naturwissenschaftlichen Forschung, Band VII. 1913.
> An excellent short summary of recent work on pearls and pearl formation.

KUNZ and STEVENSON. The book of the Pearl. New York, 1908.
> This is a large and very beautiful work. It is the best book of a general kind that has been published on the subject and covers a very wide range.

SEURAT. L'Huître Perlière. Paris (1900).
> A cheap but quite excellent summary of the subject—some sections now out of date.

Special

BOUTAN, L. Les perles fines. Leur origine réelle. *Arch. Zool. expér. gén.*, 4 sér., T. 2, 1904.
> (The origin of the pearl sac in Mytilus is described.)

DIGUET, L. Sur la formation de la perle fine chez la Meleagrina. *Comp. Rend. Acad. Paris*, T. 128, 1899.

BIBLIOGRAPHY

DUBOIS, R. Various papers on pearl formation have appeared in the *Comp. Rend. Acad. Paris*, T. 133 (1901), T. 138 (1904), T. 154 (1912).

FILIPPI, F. de. Sull' origine delle perle. Übersetzt von Küchenmeister. *Arch. f. Anat. u. Phys.* 1856.
 (One of the early interesting papers on the origin of pearls.)

HARLEY. Composition and structure of pearls. *Proc. Roy. Soc.* London. Vols. 43 and 45. 1888 and 1889.

HERDMAN, W. A. Presidential Addresses to the Linnean Society. *Proc. Linn. Soc.* London, 1905 and other years.
 (Summary of old and new theories of pearl formation and also the method of artificial pearl stimulation discovered by Linnaeus.)

—— Report on the pearl oyster fisheries of Ceylon. *Royal Society.* London, 1903—1906.
 (A report of five volumes on the Pearl banks of Ceylon, and the recommendations of the scientific advisers to the Ceylon Government. An account of pearl formation.)

JAMESON, H. L. On the Origin of Pearls. *Proc. Zool. Soc.* London, 1902.
 (An account of the origin of pearls in the edible mussel.)

—— Studies on Pearl-oysters and Pearls. *Proc. Zool. Soc.* London, 1912.
 (An account of the structure of the pearls and shell of the oriental pearl oyster.)

RUBBEL, A. Über Perlenbildung u. s. w. bei *Margaritana margaritifera*. *Zool. Jahrb.* (Anat. Abt.), 32. Bd., 1911.
 (A detailed account of some recent work on the formation of pearls in the fresh water pearl mussel of Europe.)

GLOSSARY

Abalone. A name applied to the mollusc Haliotis, or, as it is called in Jersey, the "Ormer."

Adductor muscle. A muscle passing across from one valve of a bivalve to the other, for the purpose of closing the shell (see Fig. 4).

Arthropod. A segmented invertebrate animal with a more or less hard external case and jointed limbs (for example, the Lobster or an Insect).

Baroque. Any pearl of irregular form; the name includes a large range of varieties.

Baskets. Brass sieves used in Ceylon for sifting apart pearls of different sizes.

Blister. A structure sometimes incorrectly termed a pearl, but which in reality is nothing but an excrescence of nacre attached to the shell. A deposit of nacre round an object placed between the shell of a mollusc and the mantle.

Blue pearls. Dark coloured pearls of opaque slate-blue colour.

Bombay pearls. Pearls chiefly from the oriental pearl oyster (*Margaritifera vulgaris*) which are fished in the Persian Gulf and Red Sea and exported through Bombay.

Button pearls. Dome-shaped pearls with one surface almost plane.

Byssus. The threads secreted by glands in the foot of certain shellfish, for attachment to hard bodies or to one another.

GLOSSARY

Cestode. A parasitic flatworm consisting of a hooked head (scolex) and a segmented body (proglottides). The common tapeworm is a cestode.

Cilia. Microscopic hair-like protoplasmic processes from cells, which usually have the power of vibrating rhythmically.

Cocoanut pearls. Pearls from the oyster or clam of Singapore.

Coelom. A restricted term applied to the body-cavity and other spaces in certain animals, which are formed in a distinctive manner.

Conch pearls. Pearls, often pink in colour, from the univalves Strombus and Cassis.

Conch shells. Univalves of the species Strombus and Cassis employed for making cameos.

Cyst pearls. True pearls which occur in the tissues of pearl-producing shellfish, in a pearl sac and away from the shell.

Dead pearls. Pearls with practically only a lustreless and a dead white appearance.

Dorsal. The upper side of an animal as contrasted with the lower, or ventral side. The backbone side of vertebrated animals.

Drilled pearls. Pearls bored for mounting purposes.

Drop pearls. Oval pearls or pear-shaped pearls.

Dust pearls. Very small seed pearls.

Echinodermata (prickly-skinned). A group of marine radially-symmetrical animals including the starfishes, sea urchins, sea cucumbers and featherstars.

Ectoderm. The epithelium bounding the outer surface of an animal's body, and its derivatives.

Epithelial sac. A sac composed of epithelium, as, for example, the pearl sac.

Epithelium. A layer of cells bounding a surface in the body of an animal, whether external or internal.

Fresh water pearls. Pearls from freshwater shellfish of the family Unionidae.

Haemoglobin. A red pigment found in blood, which plays an important part in the processes of respiration. Its place is usually taken by other non-red pigments in invertebrate animals.

Half pearls. Pearls sawn into halves.

Hinge pearls. Pearls of irregular elongated shapes, found near the hinge of shellfish from fresh waters.

Hypostracum. The most internal layer in a bivalve shell at the places where the adductor muscles are attached. It is composed of minute columns set at right angles to the plane of the shell (see Fig. 2).

Lamellibranchiata. The group of molluscs known popularly as bivalves and characterised by the possession of lamella-like gills.

Lingah pearls and shells. Shells and pearls from the Persian Gulf, often the same kind as the Bombay pearls.

Mantle. Two flaps arising one on either side (right and left) of the body of a bivalve. Each flap has the same shape as the valve of the shell against which it always lies. It is composed of connective tissue with an outer bounding layer of epithelium, and this epithelium on the surface in contact with the shell is responsible for both shell and pearl formation.

Mollusc. A soft-bodied, non-segmented invertebrate animal which typically possesses a hard shell. This shell may be univalve as in the snails, or bivalve as in the oyster, cockle and mussel. Sometimes it is reduced and internal, as in the slugs and cuttlefishes.

Muscle pearls. Small pearls found in the muscular tissue near its attachment to the shell.

Mytilus pearls. Pearls found occurring in the common edible mussel (Mytilus).

Nacre. The mother-of-pearl layer which lines most bivalve shells and forms all or only the outer layers of nacreous pearls.

Naked diving. Diving without special dress, etc.

GLOSSARY 143

Oriental pearls. Pearls from true pearl oysters of the genus Margaritifera; applied first to pearls from the Indian seas.

Paar. The banks or shallows in the Gulf of Manaar on which the pearl oysters live.

Pallial cavity. The space inside a bivalve shell, which is part of the outer world, and usually filled with sea water or fresh water according to the home of the mollusc.

Pearl sac. The little sac or bag of epithelium which secretes the layers that build up a pearl.

Periostracum. The outermost horny layer of mollusc shells.

Petal pearls. Flattened, leaf-like pearls.

Pinna. A bivalve, common in the Mediterranean, from which pearls are obtained. These pearls have little lustre.

Placuna. The window-pane oyster of Ceylon.

Prismatic substance. The substance of which the prismatic layer of mollusc shells is composed. It lies just under the outermost layer of the shell, and on its inner surface comes into contact with the nacreous or mother-of-pearl layer (see Fig. 2).

Seed pearls. Small pearls, round or irregular, but only $\frac{1}{4}$ grain or less in weight.

Skinning or **peeling.** The removal of the outer layer of a pearl.

Slugs. Nacreous excrescences; irregularly shaped pearly masses from fresh water shells.

Tapes. A bivalve, common off the coasts of England and France.

Trematode. A flat worm which is parasitic either externally or internally. The most familiar example is the "liver fluke" of the sheep. Another species causes pearl formation in the edible marine mussel.

True pearls. Pearls which are formed in a pearl sac, and which are found free in the shellfish, without any connection with the shell. They may not be composed of any nacre. Any, or all, layers of the shell may be found.

Twinned pearls. Pearls which have come together and then been coated with layers of nacre (see Fig. 7).

INDEX

Artificial pearls, 64, 118, 127
    stimulation as a means of forming, 66, 128

Baroche pearls, 62
Blisters, 53, 64, 66, 95

Cameos, 122
Ceylon pearl oyster, 12
    attachment ground of, 46
    classification of, 18
    development of, 42
    enemies of, 30, 50, 134
    external features and shell of, 18–25
    mantle of, 25
    muscles of, 27, 29
Colour of pearls, 5, 63
Composition of pearls, 61
Cultivation of pearls, 129

Definition of pearls, 52, 53
Drilling of pearls, 119
"Dying" of pearls, 120

Famous pearls, 120

Government commission (Herdman's), 104, 130

Inspection of pearl grounds, 73

Medieval use of pearls as medicine, 7

Nuclei of pearls, 59, 98, 111

Occurrence of pearl oysters, 12
Origin of pearls, 8, 91, 102

Paars, 46
    artificial, 44, 49, 133
Parasites of pearl molluscs, 96, 99
Pearls
    references in Classics to, 4, 5
    in edible mussel, 15, 98
    in plants, 123
    references in Scriptures to, 1, 2
Pearl fisheries
    American, 88
    Australian, 13, 85, 128
    Ceylon, 2, 10, 68, 130
    European, 7, 9, 14, 89
    Japanese, 88
    Persian, 83

Science and pearl fishing, 125
Separation of pearls, 81
Skinning of pearls, 58, 119
Stringing of pearls, 119

Theories of pearl formation, 93 *et seq.*

Valuation of pearls, 118
Value of pearl fisheries, 117, 134
Varieties of pearls, 54, 107

For EU product safety concerns, contact us at Calle de José Abascal, 56–1°,
28003 Madrid, Spain or eugpsr@cambridge.org.

www.ingramcontent.com/pod-product-compliance
Ingram Content Group UK Ltd.
Pitfield, Milton Keynes, MK11 3LW, UK
UKHW040157230326
469255UK00012B/147